T0187903

Omic Association Studies with R and Bioconductor

Omic Association Studies with R and Bioconductor

Juan R. González
Alejandro Cáceres

CRC Press
Taylor & Francis Group
Boca Raton London New York

CRC Press is an imprint of the
Taylor & Francis Group, an **informa** business

A CHAPMAN & HALL BOOK

CRC Press
Taylor & Francis Group
6000 Broken Sound Parkway NW, Suite 300
Boca Raton, FL 33487-2742

First issued in paperback 2020

© 2019 by Taylor & Francis Group, LLC
CRC Press is an imprint of Taylor & Francis Group, an Informa business

No claim to original U.S. Government works

ISBN 13: 978-0-367-72810-6 (pbk)
ISBN 13: 978-1-138-34056-5 (hbk)

This book contains information obtained from authentic and highly regarded sources. Reasonable efforts have been made to publish reliable data and information, but the author and publisher cannot assume responsibility for the validity of all materials or the consequences of their use. The authors and publishers have attempted to trace the copyright holders of all material reproduced in this publication and apologize to copyright holders if permission to publish in this form has not been obtained. If any copyright material has not been acknowledged please write and let us know so we may rectify in any future reprint.

Except as permitted under U.S. Copyright Law, no part of this book may be reprinted, reproduced, transmitted, or utilized in any form by any electronic, mechanical, or other means, now known or hereafter invented, including photocopying, microfilming, and recording, or in any information storage or retrieval system, without written permission from the publishers.

For permission to photocopy or use material electronically from this work, please access www.copyright.com (http://www.copyright.com/) or contact the Copyright Clearance Center, Inc. (CCC), 222 Rosewood Drive, Danvers, MA 01923, 978-750-8400. CCC is a not-for-profit organization that provides licenses and registration for a variety of users. For organizations that have been granted a photocopy license by the CCC, a separate system of payment has been arranged.

Trademark Notice: Product or corporate names may be trademarks or registered trademarks, and are used only for identification and explanation without intent to infringe.

**Visit the Taylor & Francis Web site at
http://www.taylorandfrancis.com**

**and the CRC Press Web site at
http://www.crcpress.com**

To our families.

Contents

Preface

The aim of the book is to offer a practical guide to researchers, graduate students and those interested in the analysis of *omic* data. While our emphasis is on the use of data in publicly available repositories, the reader interested in analyzing novel data will find settled methods for inquiring into high-dimensional biological data. We have conceived the book as a first reference to tackle specific types of data, as well as a textbook for a bioinformatics course at the MSc level. Our objective is to demonstrate how to analyze genomic, transcriptomic, epigenomic and exposomic data to explain phenotypic differences among individuals. We describe the first analyses and methods of inquiry that should be used to identify the patterns in the data that associate with a trait of interest. During the past decade numerous methods have been developed and, due to the complexity of the data, we expect many more to be devised. Nonetheless, we describe some of the most established methods that are available in the Bioconductor and R repositories, which should constitute the first line of inquiry and to which future developments should be compared against.

The methods and applications described here are all publicly available and are accessible to anyone comfortable with fitting a linear regression model in R. While we direct the reader to numerous introductory books in R and basic statistical methods, the present book is directed to users. From a basic user level, we aim to guide the readers to expand their toolkit in order to deal with *omic* data with confidence.

All the methods discussed here are part of our daily toolkit. We are regular users of all the methods and are also developers of many of them. The book is the result of compiling workshop and class material, of software package development and of years of research carried out in Juan R. González's Bioinformatics Group in Genetic Epidemiology, within ISGlobal. We have thus developed expertise in the use of the methods and in their communication, and have realized the need to offer a guide to new researchers in the field. There is a wealth of publicly available software and data, yet the landscape is overwhelming to newcomers. We offer them starting points from which to begin inquiring into the *omic* data of interest. We do not offer a complete or global view but indicate safe up-to-date entry points. As developers of some of the packages discussed, we are committed, as part of the Bioconductor community, to offer clear and reproducible documentation, clarify doubts and update new versions. We insist that packages and pipelines to assist users are also implemented so they are further improved by other developers.

The material discussed in the book is largely based on cheap high-throughput methods. They include microarrays and some sequencing methods such as RNA-sequencing. We are also aware of the developments in the collection of new high-dimensional biological data, such as Next-Generation Sequencing or those aimed at single cells. There are, however, important advantages in the use and analysis of microarrays which will keep them relevant for many years. First, association studies require cohorts and technologies to be scalable to hundreds of thousands of individuals to properly power epidemiological inferences. Microarrays clearly meet the target. While we may conceive such scalability for future sequencing, the preprocessing of data may change but the basic methods of inference would likely remain the same. In addition, microarray data is widely available and it has been an important source of continuous reanalysis to test novel focused hypotheses, confirm new results or reproduce previous findings. Finally, SNPs arrays can be additionally used to explore other genomic variants, for which specific high-throughput technology is not yet available. Therefore, association analyses in large cohorts can be performed on inversion polymorphisms and mosaicism, including the loss of chromosome Y.

This book is the result of the joint effort with other colleagues whom we have collaborated throughout the years. We would like to explicitly acknowledge and thank Carles Hernández-Ferrer, Marcos López and Carlos Ruiz who have contributed with their ideas, work and coding hours to the R packages that we have developed at the BRGE and that are discussed within the book. We are thankful to them for starting their research careers with us and for the valuable input that they have given us through their PhD projects. Roger Pique-Regi is also acknowledged for his fruitful collaboration with the R-GADA package. We would also like to thank our colleagues and collaborators from whom we continuously learn, get encouragement and intellectual stimulation. We particularly would like to mention Luis Perez-Jurado, Mariona Bustamante, Xavier Basagaña, Manolis Kogevinas, Jordi Sunyer and Martine Vrijheid, and all our colleagues from ISGlobal. We also would like to thank Tonu Esko from the Estonia Biobank for providing access to large amounts of data to test our methods, when data sharing was not a standard procedure. Finally, we would like to acknowledge support from Ministerio de Economía y Competitividad y Fondo Europeo de Desarrollo (grant number MTM2015-68140-R), Ministerio de Ciencia e innovación (grant numbers MTM2011-26515 and MTM2010-09526-E) and Ministerio de Educación y Ciencia (grant number MTM2008-02457/MTM).

The material presented in the book has been conceived as complete analysis sessions, in which initial data is available and the reader can follow, step by step, the R commands that will lead to a concrete result. Concepts and theory are introduced and explained as we go along with the analysis demonstrations. As such, all data, software and code are freely available and can be accessed and reproduced in any platform. Most data can be downloaded from the Internet. Data from the main repositories can be accessed directly within an R session, otherwise, we indicate functioning URLs at the time of publishing. Some functions have been implemented to add functionality to existing software. We have deposited them in the our GitHub repository which is publicy available at `https://github.com/isglobal-brge/book_omic_association`. Our GitHub repository (`https://github.com/isglobal-brge/`) also contains vignettes for most of the packages used in this book describing more detailed analyses. Also, the repository contains the most updated versions of the packages which include new features and bugs fixed. Specific instructions for data access and software needed are explained within each analysis demonstration.

1

Introduction

CONTENTS

1.1   Book overview

This book is concerned with the analysis of high dimensional data that is acquired at specific biological domains. The aim of the analyses is the explanation of phenotypic differences among individuals. We, therefore, search for endogenous and exogenous factors that may influence such differences. The endogenous domains on which we turn our attention are those at the molecular level involving basic DNA structure and function, which have been labeled with the *omic* suffix. In particular, we will describe current methods to analyze genomic data, which is high-dimensional at the gene (DNA sequence) level, transcriptomic data involving transcription of DNA into mRNA and epigenomic/methylomic data that relate to the epigenetic modifications of DNA. Many of the methods used at each domain overlap due to the biological nature and high dimensionality of data. However, important specificities remain, some derived from the acquisition of data and others from differences in the underlying biological processes. Within the exogenous domain, we study the high dimensional acquisition of exposure factors that are believed to influence the development or progression of individual traits.

1.2   Overview of *omic* data

Omic data refers to data collected massively in a specific *omic* domain. The notion of an unbiased scan of numerous biological entities of similar nature comes to mind. The definition is clearly operational, given the differences in understanding biological similarity and, perhaps more challenging, the varying characteristics of different biological levels. Genomic data, for instance, is ultimately concerned with the full characterization of the DNA sequence of an individual. As such, it is highly stable across tissues and the individual's lifespan. Some variations may arise in terms of somatic mutations that give rise to mosaicisms or to specific mutations, as found in tumorous cells. By contrast, transcriptomic or epigenomic data are highly variable across tissues, each of which changes on different time scales. Transcription data is highly dynamic and responds to physiological activity while epigenetic changes are expected to occur at developmental and aging rates.

   An additional consideration is the differences in the expected coverage of each *omic* data or the data's dimensionality. Nowadays, for instance, one can expect from current technology that the complete DNA sequence of an individual may be determined, or estimated to high accuracy; and therefore, genomic data is close to full coverage. However, transcriptomic data is currently far from giving us the full picture: the complete set of transcripts of an individual in a given time across all cell types. While transcriptomic data is clearly not complete, it is, however, a highly-dimensional unbiased-scan of a possible state of the transcriptome; that is, the complete set of transcripts in a biological sample of the individual.

Current extensions of *omic* data include metabolomics and proteomics, and other domains not strictly associated with specific molecular levels. These include, for instance, phenomic and exposomic data, which record multiple phenotypes and exposures at any level: molecular, organic or population. Studies including such data, therefore, allow high dimensionality on the response, traits or environmental conditions of individuals. Here, we will be concerned with studies of single phenotypes and conditions that are controlled or can be adjusted for covariates. We are primarily interested in describing subject variability on single phenotypes at a molecular level. We, therefore, study high dimensional data of DNA structure and function of groups of individuals, whose analysis methods show wide consensus. Some attention will also be given to exogenous factors given by exposomic data, which is a massive collection of environmental conditions in an unbiased manner.

1.2.1 Genomic data

The genome of an individual is the entire DNA content of all the individual's chromosomes. Genomic data comprises extensive and unbiased measurements of all the chromosomes' nucleotide sequences. Therefore, the highest possible dimensionality of genomic data is the number of nucleotides in the genome. However, it is the comparison between genomes what informs about their biological and meaningful substructures. As such, a collection of genomic data across individuals is based on the sequence variability of given structures.

1.2.1.1 Genomic SNP data

The simplest and most common structural variants in the genome are single nucleotide polymorphisms (SNPs). They are changes in only one nucleotide within a short DNA sequence that is otherwise conserved across individuals. The changes considered as SNPs are those given by only one substitution of a nucleotide for another, they are bi-allelic mutations and not rare in the population. Their allele frequencies are considered to be higher than 1%. SNPs can be detected with microarrays or sequencing techniques.

1.2.1.2 SNP arrays

Short DNA sequences, with their variant nucleotides at their ends, constitute probes that can be interrogated by its hybridization with the DNA of a given subject, which has been amplified, cut and marked with fluorescent dyes, one for each variant nucleotide or allele. Microarrays are scilico chips of millions of immobilized probes that capture the luminous DNA fragments of the subject, creating an optical pattern that is given by the individual's allele pairs, or genotypes, at each probe.

Different microarray technologies are used to genotype individuals with this approach, which is currently the most efficient and economical method to measure a substantial part of the genomic variability between individuals. The end result is an extensive coverage of SNP variants across the genomes of thousands/hundred-of-thousands of individuals. For large studies, the dimensionality of this data can achieve 10^5 (individuals) times 10^7 (SNPs), where the SNP variables are typically encoded as 0, 1 and 2 for annotated homozygous, heterozygous and variant homozygous, respectively. Annotations are complementary data on the genomic variables containing the two possible alleles at a given SNP; among adenine (A), thymine (T), cytosine (C) or guanine (C); the DNA strand, 5' to 3' (+) or 3' to 5' (-); and the alleles that should be considered as reference. Other specific considerations, that influence posterior analysis, include quality measurements of technical and biological conditions affecting SNPs and individuals.

A typical human SNP array assay includes a couple of millions of reference SNPs, from about 85 million SNPs existent in humans [23]. Neighboring SNPs are, however, highly correlated. Due to recombination, the correlation between SNPs diminishes with their distance but it is still substantial ($R^2 \sim 0.2$) for SNPs as far as 200,000 base pairs. Blocks of correlated SNPs, namely haplotypes, in reference populations have been used to impute the value of unmeasured SNPs and thus help to increase the number of SNPs of a particular study or facilitate the merging of genomic data from multiple studies [59]. The scalability of microarray-based studies is, therefore, their biggest asset to identify the likely small independent effects of numerous SNPs on complex traits [89].

SNP microarrays collect the genetic variability of individuals in known sequence variants. The known variants have been determined from reference population samples which have been fully sequenced. It remains to be determined the extent to which the selected references can offer a complete and unbiased coverage of different population samples. Despite the benefits of microarray genotyping, genome sequencing is still the ultimate source of information to fully define the genomic variability of individuals.

1.2.1.3 Sequencing methods

High-throughput sequencing methods aim to sequence all the DNA content of individuals. Broadly, in these methods, DNA is cut at small sizes (~ 100 base-pairs) or reads. Hundreds of millions of reads are then produced, which can cover the genome a number of times ($\sim 5/8$), and need to be assembled to reconstruct an individual genome. Specific sequence variants of individuals can be estimated with high accuracy. The mapping of the reads of different individual genomes to a reference genome recovers genomic SNP data with the greatest coverage, unconditioned to ancestry. The scalability of genomic data, obtained from sequencing, is, however, limited. Current technology is expensive and computationally demanding and a suitable increase in the number of individuals, required to detect the likely small effects of common variants, is at the moment unattainable.

Sequencing call of structural variants, therefore, remains an important tool to investigate rare variations and specific genomic architectures, while SNP arrays are most powerful in large studies of common genomic variation.

1.2.2 Genomic data for other structural variants

Genomic variation is rich, even between individuals with common recent ancestry. In a specific population, several DNA segments, of various lengths and up to the order of mega bases-pairs, can be found inserted, duplicated, deleted, translocated or inverted. While DNA sequencing is the best way to detect genomic variation, its price and analysis demands in large cohorts limit its use. SNP microarrays can, however, be exploited to detect many of these variants. For instance, luminous intensities used to genotype SNPs, can also be utilized to either detect regions with copy number alterations or cell populations with different genotypes (mosaicism)[167, 45]. In addition, specific haplotype patterns, which are produced by suppression of recombination, are indicative of mispairing between homologous chromosomes due to likely structural differences between them. Large and divergent haplotype groups have been associated with the suppression of recombination due to inversion polymorphisms. From genomic SNP data, inversion genotyping can be performed and their variability and functional impact can be studied in large cohorts [16].

Microarray SNP data opens the possibility to study more complex structural DNA variation in population samples across the genome. We can, therefore, exploit SNP data to have a more complete knowledge of genomic variability and to study the potential role of large structural variation in the phenotypic differences between individuals.

1.2.3 Transcriptomic data

Complex biochemical reactions are involved in the de-codification, or transcription, of DNA sequences. A direct product of these reactions is the production of RNA molecules some of which is further processed to produce proteins, the basic tools of the cells' physiology. Transcriptomic data is, therefore, a large-scale survey of the transcribed RNA repertoire of a biological sample.

The dimensionality of transcriptomic data is much smaller than that of genomic data. While in the production of a single RNA molecule, extensive and disjoint DNA may be involved, the structure of the molecule can be mapped to one gene. Genes constitute specific genomic regions of high variability in extent (up to 10^6 base-pairs) but cover in average 10,000 base-pairs of DNA sequence. In humans, the number of coding genes is estimated to be around 20,000. As such, the coding region of the genome, composed by all genes, may represent only $2 - 8\%$ of the genome. Consequently, the dimensionality of transcriptomic, or gene expression, data can widely cover a transcriptome.

Unlike the genome, there are numerous transcriptomes per individual. Given that the RNA repertoire of a cell underlies its specific functions there is at least as many transcriptomes as cell types. In addition, transcriptomes are dynamic and therefore full coverage of the conditions that alter the transcriptome is currently limited.

1.2.3.1 Microarrays

Microarrays have been extensively used to study gene expression. Messenger RNA (mRNA), RNA destined to be translated into protein, is collected from a biological sample and used to synthesize complementary DNA (cDNA). cDNA is then amplified, cut and marked with fluorescent dye and hybridized on a chip containing probes, which are DNA segments of the genes' encoding known mRNAs. Luminous patterns are analyzed to measure the content of specific parts of mRNAs, as markers for their abundance.

Each mRNA maps to a gene but one gene can be mapped by numerous mRNAs. A gene encodes different mRNA transcripts, or isoforms, that are produced by the alternative splicing of the primary RNA, the molecule directly transcribed from the DNA sequence. Therefore, transcriptomic data, obtained from hybridization of gene probes is typically a mixture of the mRNA transcripts that map to a common gene. While specific junction probes can be designed to test the abundance of a particular mRNA transcript, the complete transcriptome, given by the abundance of all genes' isoforms, has to be derived from high dimensional data of probes that cover the entire structure of all the possible mRNA transcripts. Because the coding region of a gene is given by a set of disjoint sequences, exons, it is sufficient with one probe for one exon. Therefore, a transcriptome can be inferred from exon microarray data of hundreds of thousands of probes, as many as exons in the genome. While microarray data are giving way to other sequencing-based technologies, there is a large amount of available transcriptomic data that researchers can access for re-analysis. In addition, it is economically viable for many studies.

1.2.3.2 RNA-seq

Microarrays query specific points of mRNAs requiring *a priori* knowledge. An unbiased scan of the RNA repertoire in a biological sample is clearly a sequencing of its entire RNA content. RNA-seq is the application of high throughput sequencing to RNA. Reads are mapped to the gene exons of a reference genome. As there are numerous transcripts in the sample, the production of reads in a given region is a measure of the content of the mRNAs that is encoded by the region. Therefore, the read count per exon is the main output of this type of transcriptomic data. Again, the data is not the direct observation on the transcriptome but a mixture of the gene isoforms at each exon. However, RNA-seq also provides junction reads that can inform on a particular mRNA transcript.

Compared to microarrays, RNA-seq allows detection of low expressed genes and genes with higher fold change between conditions. Furthermore, RNA-seq does not need *a priori* knowledge of probes, allows detection of genomic variants and it does not present problems like cross-hybridization. While RNA-seq is expensive and its analysis complex, transcriptomic data collection is transitioning from microarrays to RNA-seq. This is aided by the fact that expression differences between conditions (tissue or disease) are already detectable in hundreds of individuals and not in hundreds of thousands, as required by genomic studies.

1.2.4 Epigenomic data

The accessibility to DNA material is essential for the expression of genes. The way in which DNA is packed or its structure modified at a specific location can alter the fate of gene translation. For instance, the addition of a methyl group to the cytosine of a cytosine-guanine sequence 5'-3' (CpG), reduces the accessibility of DNA at this point by affecting the binding of proteins that promote transcription. DNA methylation has attracted much attention because it contributes to epigenetics (non-sequence modifications of DNA that are heritable), cell differentiation and cellular response to the environment at the genomic level [65].

Methylomic data is, therefore, the survey of the methyl modifications in the genome. Methylomic data is tissue-specific and dynamic through an organisms development and aging. Therefore, while the dimensionality of the data is bounded by the number of CpG sequences in the genome (1% in humans), a total covering of the methylomes of an individual is constrained by the number of cell types and the individual's age. CpG content is uneven across the genome and tends to concentrate in islands on the promoter regions of genes, which in humans are around 45,000 islands. Concentration on CpG islands and regions near genes can reduce the dimensionality of the data.

DNA methylation can be measured by means of DNA hybridization or sequencing. Similar to genomic and transcriptomic data, methylomic data can be obtained from microarray and high-throughput sequencing methods. In microarray-based methods, DNA material of a biological sample is firstly treated with bisulfite, which converts the unmethylated cytosines to uracils, and then amplified, which converts uracils to thymines. Therefore, methylation levels can be observed from SNP type probes of variant alleles C/T that mark the methylation status methylated/unmethylated at their genomic locations. As the collection of methylomic data is reduced to the sequencing of treated DNA, where nucleotide replacements of unmethylated cytocines by thymines are induced, high-throughput sequencing can also be applied.

Methylomic data is highly sensitive to cell type and is highly variable between individuals. Therefore, large association studies are required to observe reliable effects of methylation on phenotypes. As such, methylomic microarrays are currently favored due to their scalability to multicentric studies.

1.2.5   Exposomic data

Environmental factors are important contributors to the etiology of most complex diseases. Therefore, individual differences in multifactorial traits will not be completely explained as long as environmental conditions are not taken into account [35]. In 2005, Christopher Wild defined the *exposome* as every exposure to which an individual is subjected from conception to death [170]. While the definition parallels that of molecular *omic* data, as the total content of similar entities of a given biological domain, the totality of an individual's exposure is clearly immesurable. Efforts to bound the exposome to an operational definition need to consider both the nature of exposures and their changes over time [170]. As for the nature of exposures, one can consider those at the internal environment, specific external environment, and general external environment [171]. Whereas characterizing the time scale of the exposome is more challenging as the exposure dynamics and kinetics can change in orders of magnitude depending on the exposure. While a state of the exposome is difficult to define, following the parallels from the genome or transcriptome, a high dimensional collection of exposures can be performed, each of which is under their own spatiotemporal characteristics and methods of acquisition. Unbiased assessment of the exposome is harder to achieve as targeted exposures need to be previously defined. Higher degrees of unbiased measures can be achieved with methods that intend to scan the compounds from the exogenous origin within an organism, using mass spectrometry. However, it is still early days for such developments.

1.3   Association studies

The most immediate interest in *omic* data is to underpin trait differences between individuals at lower biological domains. Many causal relations do exist in cases were specific differences at the molecular level are amplified at a population level. For instance, Mendelian mutations in the gene *F8* can lead to hemophilia A, or de-regulation of biochemical signaling of insulin can result in diabetes. However, complex multifactorial traits typically emerge from the interactions between many units at different biological levels. Tracing back the differences between individuals at lower levels is greatly challenged by the complexity within and between biological domains.

Omic data offers an unbiased high-dimensional scan of a biological domain; and therefore, consistent patterns that associate with given individual differences in the population can be searched. In this approach, the patterns do not arise from a specific scientific hypothesis, but rather from the more general question of whether we can observe a consistent, reproducible set or arrangement of variables in a given *omic* domain that associates with subject differences. The analysis of *omic* data is therefore not based on the testing of mechanistic hypotheses. It is based on the discovery of plausible biological patterns that can guide researchers into the mechanisms. For instance, genome-wide association studies have identified numerous genomic variants associated with late-onset Alzheimer's disease [73]. Some of the variants had been associated with the disease before genomic data were available, such as the variants in *APOE*[28]. Whereas most variants are within genes, not previously associated with the disease, but offer new insights into its etiology, like the probable role of endocytosis by variants in *PICALM*[159].

1.3.1 Genome-wide association studies

Genome-wide association studies (GWASs) are based on the analysis of genomic data that try to identify SNP variants that are independently associated with differences between population samples. GWASs are any type of observational studies where specific subject differences are of interest. Specifically, the question that these studies address is if there exists *any* SNP that independently associates with subject differences. The patterns searched in this type of study are, therefore, at the univariate level. As such, massive univariate tests are performed, one for each SNP in the dataset, for which suitable inferences are drawn and tested for statistical consistency, scientific reproducibility, and biological plausibility.

A large number of GWASs have been performed in the last decade. From 2008, the GWAS catalog, sponsored by the National Human Genome Research (NHGRI) and the European Bioinformatics Institute (EMBL-EBI), has systematically collected the GWAS results of over 2,500 publications on humans [168]. These include the significant associations of SNPs for several common traits, including cardiovascular disease, cancer, type-2 diabetes and human morphology, amongst many others.

GWASs have demonstrated the ability to discover genomic variants associated with complex traits. However, they have also shown that numerous variants are needed to explain sizable amounts of common phenotypic variability [176]. In particular, for many heritable traits, there is still an amount of heritability not explained by GWASs, as the associations are typical of small in effect size [90]. In addition, the hypothesis of detecting *any* significant association amongst all the genomic variants requires strong adjustment of significance thresholds to account for random findings. The adjustment for multiple comparisons, as it is known, is in the order of the dimensionality of the genomic data, reducing statistical thresholds to orders of 10^{-8}. Therefore, studies of large sample sizes have been required to achieve enough statistical power to detect and validate findings [9]. Reasons for genomic data not accounting for the expected heritability of given traits, as measured in twin studies, are the complexity between SNP interactions or the contribution of other genetic variants, such as translocations, CNVs or inversions. Gene-environment interactions can also contribute to explain phenotype variability and lack of validation between studies. All these types of associations are also considered in specific *omic*-wide association studies.

1.3.2 Whole transcriptome profiling

While genomic data can be regarded as structural data, inasmuch as it defines the DNA content of an individual, transcriptomic data is functional. That is, it is defined by the biological states of specific tissues of an individual, at given moments. Given that some biological states are accessible to some tissues and not others, such as the neurotransmission release in the brain, or $T3$ production in thyroid, transcriptome profiling informs on the physiological functions of the tissues. As genes are tightly correlated in biochemical pathways, transcriptomic data reflects the co-regulation of genes by the correlation between their transcripts levels. Therefore, the correlation structure of transcritomic data is in itself of biological interest. Intense research is dedicated to revealing the network structure of the transcriptomic data under different conditions, such as disease state or different tissues. Numerous biochemical pathways are known and have been carefully reconstructed from well-detailed experiments [68]. Therefore, specific transcriptomic data can be methodologically compared to this pathway knowledge and inferences can be drawn on whether such-and-such pathway was active or differentiated in the data collected in a given population sample [149].

Transcriptome-wide association studies have also been extensively performed to study transcriptional differences between individuals or transcriptional signatures of phenotypes. For instance, significant and consistent differences in estrogen receptor signaling have been identified in breast cancer subtypes [29]. Given the large variability in transcriptomic data, that arises from technical (i.e. batch) and biological differences, large meta-analyses have been required to validate findings. In addition, there is an increased effort to reduce biological variability by profiling the transcriptome of single cell types.

Integration between genomic and transcriptomic data is also highly informative. Association studies in which transcription levels can be explained by genomic variation aim to determine quantitative trait loci (eQTL). These are variants that can modulate the transcription of genes by altering their coding or regulatory sequences. It has been observed that significant SNPs in GWAS are likely to be eQTLs, offering further information into the biological mechanisms underlying the associations with phenotypes [103].

1.3.3 Epigenome-wide association studies

Epigenome-wide association studies (EWAS) use epigenomic data to determine which methylated site can explain more phenotypic variability. Given that DNA methylation is affected by genomic variation and environmental exposures, the interest is to find regions in the genome that are responsive to the environment and can contribute either to adaptation or disease. Phenotypes with known heritable burden but also explained by differences in environmental exposures are of high interest. For instance, as methylation is important during development for tissue differentiation, there is interest in studying methylation patterns on head circumference, body-mass index and other developmental measures in children [131, 151]. In addition, phenotypes from mental diseases which require a gene-environment framework, such as schizophrenia or post-traumatic stress disorder [174, 160], have been approached by EWAS.

Similar to transcriptomic data, epigenomic data is strongly dependent on tissue and technical collection of data. As with GWAS, the aim of EWAS is to identify the methylation probes that are independently associated with phenotype differences between subjects. Because there are correlations between neighboring methylation probes, univariate analysis has been extended to multivariate analysis that comprises probes in extended genomic regions, which can alter the expression of a given gene or cluster of genes. Methylomic data is highly variable between individuals, therefore large meta-analyses are also considered to account for this variability and to validate findings.

1.3.4 Exposome-wide association studies

Exposome-wide association studies (ExWAS) turn their attention to the environmental risk factors as a source of phenotypic differences between individuals. Rapid developments in technology and declining costs have led to a massive increase in the amount of exposure data that can be collected for individuals over time. Current epidemiological studies are able to simultaneously measure hundreds of exposures using a combination of questionnaires, arrays of sensors and biochemical assays (see Table 1.1). Commonly assessed exposures include chemicals in the air, water, food, or household products, as well as information about individual behaviors, activities, and surrounding physical environments. Exposomic data is therefore highly heterogeneous, as it is the conjunction of different modalities that are derived from different experimental methods. The correlational structure of exposomic data Exposomic data is also dependent on the sub-types of data that are included in the exposure matrix. Therefore, given its complexity, association analyses are also performed mainly at the univariate level, in which the objective is to detect any exposure factor that is significantly associated with trait differences. Scalability and power analyses at the exposome level are difficult to establish given the heterogeneity of the exposures and the underlying mechanisms of action. Therefore, as in other *omic* studies replication is necessary to validate a finding.

TABLE 1.1
Most relevant research projects studying the *exposome* in human health.

Project	Web Site
The HELIX Project	http://www.projecthelix.eu/
The EXPOsOMICS project	http://www.exposomicsproject.eu/
HEALS	http://www.heals-eu.eu/
The Human Exposome Project	http://humanexposomeproject.com/

1.4 Publicly available resources

The ability to survey biological domains with high-throughput technology has been matched with the ability to share the data through the Internet. Collection and analysis of *omic* data are complex and can yield to not reproducible observations, particularly, if the effects are small, the biology complex and the between-study variability large.

A first objective of making the raw data of *omic* studies publicly available is to promote their independent reanalysis to increment reproducibility. Many meta-analyses have been made possible through the access of the data. However, it is becoming increasingly the important use of data repositories to confirm specific hypothesis, to find supporting evidence of initial findings at different biological domains or to test new analysis methods that adapt better to biological complexity. There are numerous data repositories, here we cite some of the most widely used with a particular emphasis on association studies. All data available through these repositories are susceptible to be analyzed using the methods described in this book.

1.4.1 dbGaP

The database of genotypes and phenotypes (dbGaP) is a public repository of genomic, epigenomic, somatic mutations, transcriptomic and microbiomic data, with associated phenotypes [88]. The repository is provided by the National Center for Biotechnology Information (NCBI). At the time of the publication of the book, the repository contained assay data for over 1.6 million of SNP arrays (2.3 hundred imputed), 10 thousand expression arrays and 10 thousand methylation arrays. It also contains high-throughput sequencing assays for 150 thousand whole exome sequencing, 50 thousand whole genome sequencing, and 25 thousand RNA-seq.

dbGaP offers open-access data and controlled-access data. Open data can be accessed without permission and pertains to general data about the study, including some phenotypic variables and summary results. dbGaP controls access to de-identified genotypes and phenotypes. Formal requests to use the data are required to ensure its use for scientific purposes, to comply with the ethical standards of the studies and to warrant proper use of sensitive data.

dbGaP is, therefore, a primary source of *omic* data and its influence is only expected to grow, as specific studies will be required to contrast their results with published raw data, and new methodologies will be able to access large sets of observations to assess their viability.

1.4.2   EGA

The European Genome-phenome Archive (EGA) is a permanent archive that promotes the distribution and sharing of genetic and phenotypic data consented for specific approved uses but not fully open, public distribution. It enables collaboration and data sharing of individual patient-level genomic and phenotype data through a controlled-access system. The EGA includes data collections for human genetics research [74].

The repository contains raw data from DNA sequencing and array-based genotyping applications, e.g. gene expression experiments, transcriptomics, epigenomics, sequencing or proteomics assays. It has processed datatypes such as genotypes, structural variations or whole genome sequence. Phenotypic data is also available, all consented for research purposes.

The archive is used as the repository of large genomic studies that include the International Cancer Genome Consortium (ICGC), the International Human Epigenome Consortium (IHEC), the The International Human Microbiome Consortium (IHMC), the UK10K project for Rare Genetic Variants in Health and Disease or the Deciphering Developmental Disorders (DDD) project among others.

1.4.3   GEO

The Genome Expression Omnibus (GEO) is a data repository specialized in functional genomic studies, including transcriptomic data from microarray and RNA-seq[8]. GEO is hosted by NCBI and has a number of on-line analysis results for specific datasets. An important advantage of GEO is that its data can be directly retrieved with Bioconductor packages in R. It hosts more than 90,000 accession entries. Most of the entries are for expression microarrays (50,000) but fastest growth of submissions are for high-throughput sequencing data (15,000). Large meta-analyses, including numerous studies of common phenotypes, can be routinely performed, such as those for breast cancer or Alzheimer's disease. Similar to dbGaP, GEO constitutes an archive of raw data for studies to be continuously consulted to advance understanding of trait differences at lower biological domains.

1.4.4 1000 Genomes

Other public data resources correspond to large multi-centric studies that have ambitiously collected data to characterize *omic* data across conditions of great interest [23]. The 1000 Genomes study is, for instance, based on the characterization of human genomes across numerous ancestries. The idea was to create a detailed map of human genomic variability, offering a platform to further support research in genetics, medicine, bioinformatics, pharmacology, and biochemistry.

The 1000 Genomes comprises genomic data from the sequencing of 2,504 individuals from 26 different ancestries with 4 × genome coverage that allows detection of variants with more than 1% frequency. This genomic data is, therefore, a reference panel of populations to impute SNPs that have not been genotyped in specific studies and to help merge genomic data across multiple studies. High dimensional meta-analysis of GWAS can be thus performed on 85 million SNPs and over 5,000 haplotypes. A subset of 423 subjects from 4 European and one African ancestry was selected for transcriptomic data collection using RNA-seq. The data was produced by the GUEVADIS project and is also freely available. The aim of GUEVADIS was, in particular, to study the transcriptome variability, in the lymphoblastoid cell line, across human populations. Focused on European ancestry, the study has shown that significant SNPs in GWASs are likely to be eQTLs, demonstrating that the integration of transcriptomic and genomic data can reveal causal variants and biological mechanisms of diseases.

1.4.5 GTEx

The genotype tissue-expression (GTEx) project aimed to study transcriptome variability across 53 different human tissues [85]. The project collected transcriptomic data for 714 donors, 635 of which were also genotyped, which allows the study of specific changes in the relationship between genomic and transcriptomic data across tissues. Genotype data were collected with SNP microarrays covering 5 million SNPs, while transcriptomic data was obtained from RNA-seq, with 50 million aligned reads per sample. eQTL analyses have been performed and its results are available through a web-browser. Specific queries can be performed that inform on the SNPs that modulate gene expression and splicing across tissues. The integration of genomic and transcriptomic data through eQLT analysis aims to guide GWAS results into the mechanisms underlying the associations between SNPs and phenotypes.

Data is freely available, as the GTEx project intended to offer a resource to find further support of novel findings, develop new methods for integration of genomic and transcriptomic data and investigate the variability of transcription across tissues.

1.4.6 TCGA

The cancer genome atlas (TCGA) is an initiative to collect multiomic data to support cancer research [156]. It is sponsored by the national cancer institute (NCI) and the NHGRI. The project has collected data on 33 different types of tumors in 11 thousand patients. *Omic* data includes high-throughput DNA and RNA sequencing , SNP, DNA methylation and reverse-protein arrays. It has generated 2.5 petabytes of information by its closure in 2017. However, further initiatives plan to build on this initial effort.

The objective of the project was to support research aimed at assessing the extent to which multiomic variability can explain differences between individuals in cancer susceptibility, cancer types, progression and treatment. As such, the data has supported, at the time of publication of this book, more than a thousand studies, including those by independent researchers from the TCGA network. Clearly, the analysis of such massive data to account for biological complexity across different biological domains will take decades to complete.

1.4.7 Others

There are several Biobanks that also provide free access to different *omic* data. The UK Biobank (UKB) is a prospective cohort study with deep genetic, physical and health data collected on 500,000 individuals across the United Kingdom from 2006-2010 (`https://www.ukbiobank.ac.uk/`). The Estonian Biobank contains genetic information about 50,000 individuals from the Estonian population as well as data from different resources including medical records (`https://www.geenivaramu.ee/en`). The BioBank Japan project has a registry of patients diagnosed with any of 47 common diseases and genomic data of 200,000 patients.

ReCount is a specialized repository for RNA-seq data. It stores processed and summarized expression data for nearly 70,000 human RNA-seq samples `http://bowtie-bio.sourceforge.net/recount`. The data, accessible through a web-application (`https://jhubiostatistics.shinyapps.io/recount/` and the Bioconductor's package *recount* [22], is made available with the aim of reproducing the expression profiles of reported findings. For instance, the RNA-seq data from the GTEx project is stored in ReCount. There are several Bioconductor data packages including *omic* data. Among others, the package *curatedTCGA* contains different objects corresponding to TCGA tumors that integrates RNA-seq, copy number, mutation, microRNA, protein with clinical/pathological data.

A source of publicly available exposomic data is offered by the National Health and Nutrition Examination Survey (NHANES) `https://www.cdc.gov/nchs/nhanes/index.htm` [107]. NHANES is a US national survey that covers demographics, health, nutrition, and environmental chemical exposures. NHANES started surveying in 1999, repeating every two-year cycles. The data is also available through R by the package *RNHANES*.

1.5   Bioconductor

Collection of high dimensional data at different biological domains demands the development of new analysis methods and generalizations of old ones. There are multiple ways in which data can be stored, preprocessed and analyzed. Diversity arises from technical capabilities, experimental conditions, and scientific hypotheses to be tested. Therefore, as volume and complexity of data increases so analytical methodology does. A proliferation of "in-house" software to address the specific needs of studies greatly challenged reproducibility, absorption and further development of the methods by a wide community of users. Bioconductor is an open source software project aimed to address these issues, by orchestrating software development in the programing language R to analyze high-throughput biological data [40].

1.5.1   R

Parallel and independently from the development of *omic* data acquisition, the data analyst community made important advancements into the integration and sharing of methodologies through the extensions of R. R is a high-level programing language that initially focused on the implementations of functions for statistical analysis. From the developer's perspective, R offers flexible syntax for object-oriented programming, which is easily packaged into software units with clear functionality. As free software, package contents can be modified by other developers or incorporated into other packages. The great flexibility of R has promoted software development in diverse research fields, in particular, those that need to quickly integrate new statistical analyses and visualization methods. R packages are shared through various public repositories such as The Comprehensive R Archive Network (CRAN), GitHub and Bioconductor.

In the production of R packages, great effort is put into documentation that includes detailed information on how to use its functions, example data, and demonstrations. In addition, manuals in form of vignettes are distributed to guide users into the specific tasks supported by the packages. The code in the examples and in the vignettes must be reproducible in any platform and by any R user. Initiatives to increase the reproducibility of code and reporting have been also incorporated into the production of packages. Comprehensive information of software, clearly delimited to achieve concrete tasks, have greatly incremented the users base of R to non-developer analysts. The R user's community is highly active on the Internet and packages are typically found by common Internet queries on specific topics of interest. In its website, CRAN offers the list of the packages available, all of which can be installed in R using the command

```
> install.package("nameOfPackage")
```

once the package is installed, it is accessible in each R session by

```
> library("nameOfPackage")
```

1.5.2 *Omic* data in Bioconductor

Bioconductor offers methods to analyze a wide range of *omic* data and to access publicly available resources [63]. Important additional tools include annotation resources and visualization capabilities. Bioconductor's project integrates all these facilities under common data structures that enable easy integration of new data and methods into existing workflows. Users can find sufficient information at different levels, such as functions, packages, and workflows, which allows them to combine and develop analysis strategies with specific needs. Bioconductor's packages are continuously growing, as technologies evolve and produce new data, and developers create packages with new capabilities. However, numerous packages have been settled into standard procedures, through an intense use and feedback, from user to developer, that continuously put to test the underlying methods. In particular, Bioconductor's methods to preprocess and perform association studies of genomic, epigenomic and transcriptomic data have achieved great consensus. They include microarray and high-throughput sequencing data.

Integration between analysis methods and retrieval of data from repositories is highly coordinated. Specific packages have been developed to use R packages to query specific databases. For instance, GEO, can be queried with the package *GEOquerry*. The package has implemented the function `getGEO` to download data of a study with a given accession number from the GEO website. Data is retrieved and made available in R as a variable of class `ExpressionSet`, a recognizable data structure of Bioconductor. The data can thus be analyzed following established workflows, to reproduce the study's reported results, or to further exploit the data to test alternative hypotheses or methods. Specific projects like TCGA and GTEx have also R packages, supported by Bioconductor, to retrieve their data.

Previously, Bioconductor's installations used the commands

```
> source("https://bioconductor.org/biocLite.R")
> biocLite("nameOfPackage")
```

Currenlty, packages in Bioconductor, like *GEOquery*, are installed with the library *BiocManager*

```
> library(BiocManager)
> install("GEOquery")
```

Once installed, the package can be loaded as an usual R package

```
> library("GEOquery")
```

GitHub `https://github.com/` is another R repository, mainly used to deposit development version of new packages. Packages from GitHub can be installed into R using the *devtools* package

```
> library(devtools)
> install_github("nameOfRepository/nameOfPackage")
```

For instance, the author's GitHub repository is `isglobal-brge`, from which multiple packages discussed in the book can be installed

```
> library(devtools)
> install_github("isglobal-brge/nameOfPackage")
```

1.6   Book's outline

The book is designed to give an introduction on how to use established tools to analyze association studies of genomic, transcriptomic, methylomic and exposomic data. We focus our discussion on publicly available datasets. The aim for this is double: to help readers who are interested in acquiring the analytical tools to start analyzing real datasets and to show those working on statistical methodologies how to access a large amount of biological data. In addition, readers who are interested in learning how to exploit available data can start proving specific hypotheses, or find further support for specific results. We, therefore, start in Chapter 2 with case studies, whose main objective is to illustrate how to access particular data repositories and how to obtain the main results that can be expected from a standard analysis. Chapter 3 describes how to deal with *omic* data in Bioconductor. Chapters 4 to 9 are dedicated to explaining in detail the preprocessing and analysis methods, functions and visualization tools of genomic, transcriptomic, epigenomic and exposomic association studies. Chapter 10 gives a first approach to the integration of *omic* data with biochemical pathways, through enrichment analysis methods. Chapter11 describes integration between different *omic* data-sets including how to gather results into functional, disease and pathway annotations and how to perform multi-*omic* data analysis using advanced multivariate methods.

All data in the book is freely available. Most of it can be directly downloaded from the public repositories however some data has been compiled in an R package to explain specific analyses. Those data are available at `https://www.github.com/isglobal-brge` and can be installed with

```
> library(devtools)
> install_github("isglobal-brge/brgedata")
```

The data is loaded in R with the command

```
> library(brgedata)
```

The *brgedata* package contains several files in specific formats (binary or text files) that will be used throughout the book. Data stored in file format are found in the folder **extdata** that is created when installing the package and can be accessed by

```
> path <- system.file("extdata", package="brgedata")
```

Binary data are accessed by using the **data** function. For instance, data of a SNP association study can be retrieved into R by:

```
> data(asthma, package = "brgedata")
> asthma[1:5, 1:10]
   country  gender      age      bmi smoke casecontrol rs4490198 rs4849332
1  Germany   Males 42.80630 20.14797     1           0        GG        TT
2  Germany   Males 50.22861 24.69136     0           0        GG        GT
3  Germany   Males 46.68857 27.73230     0           0        GG        TT
4  Germany Females 47.86311 33.33187     0           0        AG        GT
5  Germany Females 48.44079 25.23634     0           1        AG        GG
  rs1367179 rs11123242
1        GC         CT
2        GC         CT
3        GC         CT
4        GG         CC
5        GG         CC
```

The package contains the following datasets:

```
> library(brgedata)
> ls("package:brgedata")
 [1] "asthma"        "breastMulti"     "breastMulti_list"
 [4] "brge_expo"     "brge_gexp"       "brge_methy"
 [7] "brge_prot"     "genesAD"         "gwascatalog"
[10] "lusc"
```

2

Case examples

CONTENTS

2.1 Chapter overview

In this chapter, we will show, with five case examples, how to retrieve data from five public repositories and some basic analysis that can be performed on the data. We introduce the functions and packages in R/Bioconductor that can be used to perform specific queries, retrieval, and analysis, all within a single R session. Further chapters will treat in detail the packages and the functions used to produce the results.

2.2 Reproducibility: The case for public data repositories

Accessibility to primary data has been strongly motivated by the research community to encourage reproducibility of results. Studies based on *omic* data collection are particularly sensitive to reproducibility issues due to the variety of methods and strategies of analyses, and the numerous small effects and complex interactions that may underlie a particular pattern in the data.

For a given dataset and a given analysis strategy, it is at least expected that the results obtained by two different analysts will be the same. This level of reproducibility is analytical and can be tested with independent analyses of one reference study. This is easily achieved when primary research data is freely available for reanalysis and the methods together with their implementations are clearly explained and documented [147]. The second level of reproducibility refers to the validity of a scientific observation. In this case, we expect that under one analysis strategy the pattern observed in one study is reproduced in another independent study. Having the data of independent studies freely available clearly motivates validation of results.

In addition to reproducibility, the access to primary data has motivated the testing of novel methodologies. In this case, two different methodologies can be tested on the same dataset and study their differences and commonalities.

The use of freely available data has greatly contributed to advance reproducible research in studies with high dimensional data. As such, the initiatives of sharing data together with strengthening public repositories are only going to increase and become a common practice in future research programs.

2.3   Case 1: dbGaP

dbGaP is a data repository of primary research on genome-wide association studies. The detailed description of the studies available can be queried in https://www.ncbi.nlm.nih.gov/gap. Studies can be queried by keywords, i.e. "Alzheimer" or directly by its accession number. Our first case example is based on the summarized data of a study on late-onset Alzheimer's disease (LOAD). The NIA-LOAD study was carried out by the National Institute of Aging (NIA) on families with at least two affected siblings and unrelated controls [77]. The database contains 5,273 individuals with genomic data (SNP microarray), and 5,220 individuals with phenotypic information. The dbGap's accession number is phs000168.v2.p2, a full description of the study can be found in the dbGap's website, corresponding to the accession number.

Data is available under controlled-access. Authorization for use of genotype data needs to be granted by the NIH Data Access Committee (DAC), who evaluate the purpose of use and handling of data by researchers and institutions. Research and dissemination of results are encouraged with proper acknowledging of the study. Interested readers should apply for data access. Note that this is the only example where data access is controlled, all other data in the book is unrestricted. Here, we illustrate how to display *summarized* results of reported GWAS of the LOAD-NIA study. The phenotypic variables are distributed in the dataset LOAD610K_Subject_Phenotypes. In particular, they report the first four principal components (PC) of SNP array data, comprising 599,011 SNPs and 3,007 subjects. A PC analysis of genomic data is used to determine the components that capture most genetic variability between the subjects. Therefore, individuals represented in the first PC components will be clustered in groups according to similar genetic background. Clearly, ancestry is the strongest predictor of common genetic background and therefore PC analysis is used to infer ancestral similarities and differences between individuals, based on the genomic data. Therefore, one can expect that there is a strong correlation between the first PC of the genomic data and the self-reported ancestry. A scatter plot between the first two PCs of the LOAD-NIA genomic data (encoded in variables `AllEthnicity_PC1` and `AllEthnicity_PC2`), colored by the self-reported ancestry (encoded in `Race`) clearly illustrates this point.

Data is loaded in R as it is distributed in dbGap, in the appropriate phenotype subdirectory of the complete `LOAD610K_Subject_Phenotypes` dataset.

```
> data <- read.delim(
+ file = "phs000168.v2.pht000707.v2.p2.c1.LOAD610K_Subject_Phenotypes.GRU.txt",
+ comment.char = "#")
>
> names(data)[1:18]
 [1] "dbGaP_Subject_ID"   "SUBJ_NO"           "SEX"
 [4] "Dx_Level"           "Case_Control"      "RecruitedAsControl"
 [7] "ConType"            "BirthYr"           "AgeAtLastEval"
[10] "VitalSt"            "AgeDeath"          "Autopsy"
[13] "Race"               "Hispanic"          "AllEthnicity_PC1"
[16] "AllEthnicity_PC2"   "AllEthnicity_PC3"  "AllEthnicity_PC4"
```

We obtain the self-reported **race** variable as described in the variable report documentation

```
> race <- as.factor(data$Race)
> table(race)
race
   1    2    3    4   50   99
2692  123    3    3   12    1
```

We can then plot the first two PCs in the dataset and color them according to **race**.

```
> mycols <- c("gray90", "black", "gray70", "gray50", "gray20", "white")
> cols <- as.character(factor(race, labels=mycols))
> plot(data$AllEthnicity_PC1, data$AllEthnicity_PC2,
+      type="n", main="PCA LOAD-NIA",
+      xlab="PC1", ylab="PC2")
> points(data$AllEthnicity_PC1, data$AllEthnicity_PC2,
+      col = cols, pch=0:5)
>
> legend("bottomright", c("white", "black", "american indian",
+                          "asian", "other", "missing"),
+      col=mycols, pch=0:5)
```

PC components are strong predictors of ancestry and relevant measures of population stratification. Therefore, they can be used to identify individuals with strong genomic differences from a given population sample. In addition, they are important covariates to account for in association studies of genetic variants, as we will discuss in Chapter 4.

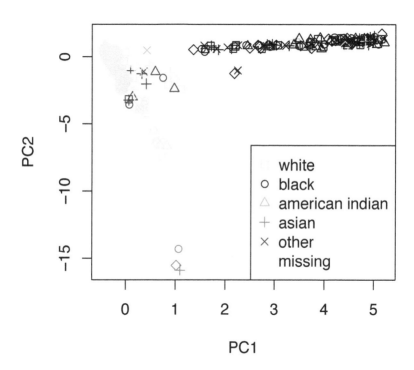

FIGURE 2.1
Genome-wide PCA of LOAD-NIA dataset.

A LOAD-GWAS on 1,877 unrelated individuals was performed and its re-sults stored in the `Run_precompute_1877_unrelated_samples.assoc` dataset, under quality control data. This can be considered as a preliminary result, given that the aim of the study is the familial segregation of the disease. We can load these results to explore them

```
> results <- read.table(
+           file="Run_precompute_1877_unrelated_samples.assoc",
+           header=TRUE)
> head(results)
  CHR      SNP BP A1     F_A     F_U A2    CHISQ      P     OR
1   0 MitoT217C 217  C 0.01410 0.01131  T 0.5753000 0.4481 1.2500
2   0 MitoG228A 228  A 0.05242 0.05436  G 0.0697000 0.7918 0.9623
3   0 MitoG247A 247  0     NA      NA  0      NA     NA     NA
4   0 MitoC295T 295  T 0.09828 0.09215  C 0.4046000 0.5247 1.0740
5   0 MitoC458T 458  T 0.05438 0.04881  C 0.5907000 0.4421 1.1210
6   0 MitoC464T 464  T 0.07411 0.07420  C 0.0001076 0.9917 0.9987
```

As we can see the association between each SNP in the genomic data and the status of the disease (case/control) was tested with a χ^2 test, which will be described in Chapter 4. The genomic location of the variant (chromosome, name of the variant, and base-pair position), the strength of the association (χ^2) and the P-value are also reported. We can check that associations strengths across all the SNPs are consistent with a χ^2 distribution. A quantile-quantile (Q-Q) plot between the χ^2 observed estimates and the quantiles of an actual χ^2 distribution should reproduce the identity line ($y = x$) when no association is expected, and the null hypothesis is not rejected. For most SNPs, this should be the case. For association with unusually large values of the observed χ^2, we expect that if they are true findings, they strongly depart from the identity line. In those cases, we reject the null hypothesis. For LOD-NIA data, we can extract the observed χ^2 and produce a Q-Q plot from the `qq.chisq` of the Bioconductor's *snpStats* package.

```
> library(snpStats)
> qq.chisq(results$CHISQ)


           N      omitted       lambda
5.81348e+05 0.00000e+00 1.03446e+00
```

The points in the Q-Q plot mainly fall on the identity line, showing lit-tle inflation or depletion of genomic P-values. There are extreme χ^2 esti-mates, though, that depart from the expected distribution. These are can-didate findings. We can inspect which P-values are lower than any sin-gle significant finding expected by chance within 599,011 tests (number of SNPs in the data set). At least one false significant finding will be likely for $P = 0.05/599,011 \sim 8.3 \times 10^{-8}$. Therefore, unlikely findings by chance are those SNPs showing a $P < 8.3 \times 10^{-8}$. Such lowering of the significance threshold is therefore necessary to correct for multiple testing. The particular

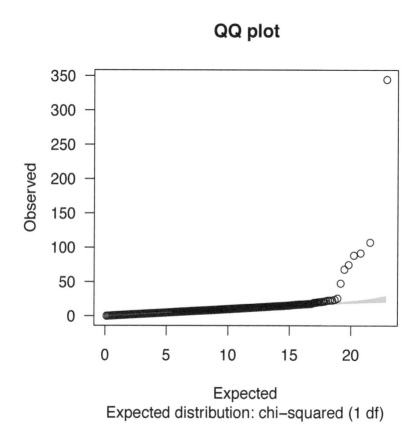

FIGURE 2.2
QQ plot LOAD-NIA.

adjustment of the *P*-values by the number of tests is a Bonferroni correction. We can perform the adjustment with `p.adjust`. In the LOAD-NIA dataset, we observe that 7 SNPs are reported with *P*-values lower than chance, after Bonferroni correction

```
> pval <- results[, c("SNP", "CHR", "BP", "OR", "P" )]
> selSig <- p.adjust(pval$P, method="bonf")
> subset(pval, selSig<0.05)
                 SNP CHR        BP     OR          P
547143 rs10402271   19 50021054 1.6000 4.596e-12
547157      rs6859   19 50073874 1.8600 5.210e-21
547160    rs157580   19 50087106 0.4759 3.709e-25
547161   rs2075650   19 50087459 4.5780 6.285e-77
547162   rs8106922   19 50093506 0.5626 1.517e-16
547163    rs405509   19 50100676 0.5308 1.026e-21
547166    rs439401   19 50106291 0.5395 5.319e-18
```

For each SNP in the genomic dataset, we can plot its *P*-value $(-log(P))$ against the chromosome position in what is called a Manhattan plot. Re-formatting the GWAS results allows the use of the plotting function `manhattanPlot` from our repository (https://github.com/isglobal-brge /book_omic_association/tree/master/R) that uses *tidyverse*, *ggplot2* and *ggrepel* packages (Figure 2.3). Other option is to use the function `manhattan` from the package *qqman* which requires the same input.

```
> library(tidyverse)
> library(ggplot2)
> library(ggrepel)
>
> # remove missing P-values
> pval.nona <- pval[!is.na(pval), ]
> # select SNPs with p-value lower than 0.01
> # to speed up the plot without loosing information
> pval.nonasig <- pval.nona[pval.nona$P<1e-2, ]
>
> manhattanPlot(pval.nonasig, color=c("gray90", "gray40"))
```

Note that all the SNPs fall within the same genomic region that spans the *TOMM40* and *APOE* genes in chromosome 19. Genomic variability of this region has been extensively validated in several independent studies [73]. We can plot the *P*-values in the region of interest using *LocusZoom* (http: //locuszoom.org/) as shown in Figure 2.3.

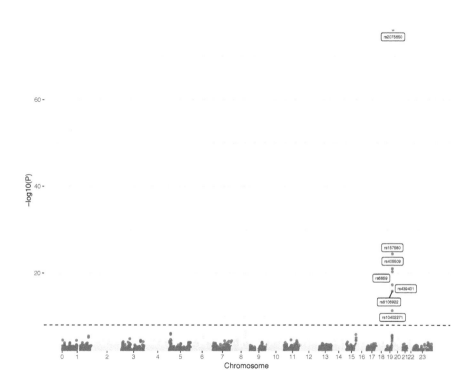

FIGURE 2.3

GWAS results on unrelated individuals of the LOAD-NIA study. Each point is a SNP in the genomic location described in the x axis. On the left y axis, the $-log_{10}(P)$ is plotted. The dashed line indicates significance threshold that corrects from multiple testing (i.e. genome-wide significance level).

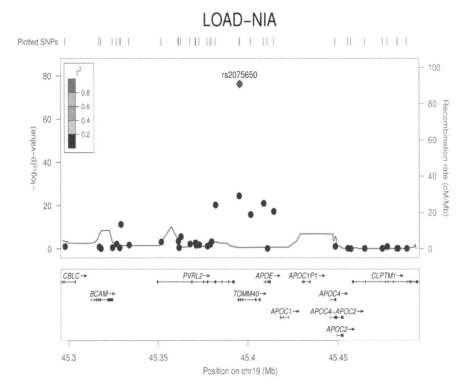

FIGURE 2.4

LocusZoom plot of significant GWAS results on unrelated individuals of the LOAD-NIA study. Each point is a SNP in the genomic location described in the x axis, where genes are represented. On the left y axis, the $-log_10(P)$ is plotted, on the right the recombination rate is given.

2.4   Case 2: GEO

Association analyses of transcriptomic data for case/control studies are performed to describe the differences in transcriptomic patterns between the groups of subjects. This type of study will be covered in detail in Chapter 7. Transcriptomic data can also be used to test specific hypotheses derived from GWASs. GWASs provide results that in most cases are to be interpreted mechanistically. The hypothesis of finding *any* significant association is of statistical nature, as it is not motivated by existing knowledge of a trait. As such, GWASs' scientific interpretation is *a posteriori*. It is, therefore, important to put their results into a wider context, in particular, within other biological levels. Further evidence of meaningful associations can be searched in transcriptomic studies, in which differences in the transcription level of a hypothesized gene may be correlated with trait differences. Alternatively, if transcription levels are available in the general population one can ask if the SNPs associated with a trait modulate the expression of the gene as eQTLs.

As an example of hypothesis testing using trascriptomic data, we analyze the data of a case-control transcriptomic study of Alzheimer's disease, matched by age and sex [142]. The data of the study is available in GEO (https://www.ncbi.nlm.nih.gov/geo/), with accession number GSE63061. Our aim in the analysis is to test whether genes with SNP variants found significantly associated with Alzheimer's disease, in the LOAD-NIA GWAS, show transcription differences between cases and controls. We thus intend to test the plausibility that the genes found with genetic variants associated with LOAD are also deregulated in disease and, therefore, establish a link between gene structure and function. Additionally, we can test the deregulation of a candidate gene for the development of Alzheimer's disease. *APP* is a gene involved in the pathology of the disease as it encodes the amyloid-β peptide whose accumulation in the brain in plaques is one of the diagnosis criteria in postmortem samples.

Data for specific accessions in GEO can be downloaded in R using the getGEO of the Bioconductor's package *GEOquery*.

```
> library(GEOquery)
> gsm.expr <- getGEO("GSE63061", destdir = ".")
> gsm.expr <- gsm.expr[[1]]
```

This function downloads a data file called GSE63061_series_matrix.txt.gz that is retrieved automatically in an ExpressionSet object (see Chapter 3 for a description of these type of objects) that we have named gms.expr

```
> show(gsm.expr)
ExpressionSet (storageMode: lockedEnvironment)
assayData: 32049 features, 388 samples
  element names: exprs
```

```
protocolData: none
phenoData
  sampleNames: GSM1539409 GSM1539410 ... GSM1539796 (388 total)
  varLabels: title geo_accession ... tissue:ch1 (40 total)
  varMetadata: labelDescription
featureData
  featureNames: ILMN_1343291 ILMN_1343295 ... ILMN_3311190 (32049
    total)
  fvarLabels: ID Species ... GB_ACC (30 total)
  fvarMetadata: Column Description labelDescription
experimentData: use 'experimentData(object)'
Annotation: GPL10558
```

The data stores different tables including the expression profiles for each probe and subject (**assayData**), phenotype data with traits measurements and covariates of interest (**PhenoData**), and feature data with information about the probe's used in the expression array (**PhenoData**). Specific data is retrieved using the necessary functions. In particular, **exprs** and **phenoData** extract data tables for subjects' expression levels and phenotypes respectively.

```
>
> #get transcriptomic data and apply log2
> expr <- log2(exprs(gsm.expr))
> dim(expr)
[1] 32049    388
>
> #get phenotype data
> pheno <- pData(phenoData(gsm.expr))
> status <- pheno$characteristics_ch1
> status <- gsub("status: ","", as.character(status))
> table(status)
status
          AD borderline MCI          CTL     CTL to AD          MCI
         139              3          134             1          109
    MCI to CTL          OTHER
             1              1
>
> #create case control variable from AD and CLT labels
> selcaco <- status%in%c("AD","CTL")
> caco <- rep(NA, length(status))
> caco[status=="AD"] <- 1
> caco[status=="CTL"] <- 0
```

We therefore have 32049 probes for 388 individuals, 139 of which are cases and 134 are controls. The probes names are Illumina names of the form *ILMN_*** but whose mapping to specific genes and genomic coordinates can be retrieved from the feature data using the function **fData**. We locate four transcripts of interest, two for *APP*, as it is the candidate gene for Alzheimer's disease, and one for *TOMM40* and another for *APOC1*, as they were found with significant SNPs association in the LOAD-NIA study. Note that this dataset does not contain transcripts for *APOE*,

```
> #locate rows in expression data for transcripts in APP, TOMM40 and APOC1
> genesIDs <- fData(gsm.expr)[13]
> genesIDs <- as.character(unlist(genesIDs))
>
> selAPP <- which(genesIDs%in%c("APP"))[2:3]
> selTOMM40 <- which(genesIDs%in%c("TOMM40"))
> selAPOC1 <- which(genesIDs%in%c("APOC1"))
>
> selTranscripts <- c(selAPP, selTOMM40, selAPOC1)
> labTranscripts <- c("APP", "APP", "TOMM40", "APOC")
```

We test the association between case/control status and the transcripts of interest using a logistic model from the basic R function glm with option family="binomial". Chapter 7 will describe these methods in further detail. We select the *P*-values and plot the transcript distributions in violin plots from the *vioplot* package.

```
> library(vioplot)
> par(mfrow=c(2,2), mar=c(2,4,2,1))
>
> for(trans in 1:4){
+    x <- selTranscripts[trans]
+    exprsel <- expr[x,]
+
+    #association
+    mod <- glm(caco ~ exprsel, family="binomial")
+    pval <- summary(mod)$coeff[2,4]
+
+    #plot
+    lab <- paste0(labTranscripts[trans], "\n P = ",
+              as.character(round(pval,3)))
+    vioplot(exprsel[which(caco==0)], exprsel[which(caco==1)],
+            col="gray80", names=c("controls", "cases"))
+    title(lab, cex.main=0.8)
+    }
```

In our example, we see that both transcript expressions of *APP* are significantly associated with case/control status. Consistent with our hypothesis, as more amyloid-β yield is expected in cases, we observe that they appear to express more *APP* than controls. However, for the genes with genomic association with the disease *TOMM40* and *APOC1*, we do not observe significant transcription differences between cases and controls. Note that it is *APOE*'s polymorphisms that have been extensively associated with the disease. A likely interpretation of our GWAS findings is that the SNPs in *TOMM40* and *APOC1* are not causal, being part of an extended linkage disequilibrium region with the SNPs in *APOE*.

GEO is also a repository for epigenomic data. Therefore, we can further investigate the methylation differences in Alzheimer's cases and controls using publicly available data. The study with accession number GSE80970, is a case/control ($N = 148/138$) study of epigenomic data, measured with the Illumina Infinium Human 450K Methylation Array. Data is retrieved directly

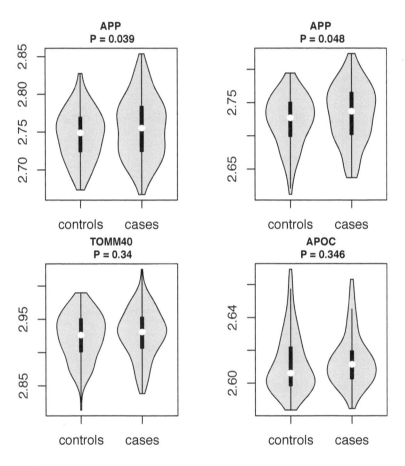

FIGURE 2.5

Transcript differences in APP, TOMM40 and APOC between Alzheimer's cases and controls

from the GEO website with the accession number. In this case, methylation levels are stored in the expression data table that can be retrieved with `exprs` as in the expression example. From phenotype data, we obtain the case/control variable that we transform into a binary variable, as the outcome for a logistic regression. In addition, the available CpG probes in *APP* are located to perform the association test. Note that the data manipulation is specific to the data version available in GEO. Here, we illustrate the extraction and manipulation of the GSE80970 accession update corresponding to November 07, 2017. Data can be downloaded from GEO by using a similar code to the one we previously used for transcriptomic data

```
> gsm.meth <- getGEO("GSE80970", destdir = ".")
> gsm.meth <- gsm.meth[[1]]
```

Now, let us illustrate how to access the required data to perform some simple analyses

```
> met <- log2(exprs(gsm.meth))
>
> #get phenotype data
> pheno <- pData(phenoData(gsm.meth))
> status <- pheno$characteristics_ch1.1
> status <- gsub("disease status: ","", as.character(status))
> table(status)
status
Alzheimer's disease              control
                148                  138
>
> #create case control variable from Alzheimer's disease and control labels
> caco <- as.numeric(status=="Alzheimer's disease")
>
> #locate CpGs in APP
> genesIDs <- fData(gsm.meth)$UCSC_RefGene_Name
> genesIDs <- as.character(unlist(genesIDs))
>
> selAPP <- grep("APP;APP",genesIDs)
> length(selAPP)
[1] 21
```

We are interested in performing logistic regressions of case/control status with the methylation levels of 21 CpGs in *APP*. We extract the P-values of association and test which of those would have been unlikely found by chance in the number of tests performed. We thus adjust for multiple comparisons considering significant associations with those P-values $< 0.05/21 = 2.3 \times 10^{-3}$

```
> pval <- sapply(selAPP, function(probes){
+     mod <- glm(caco ~ met[probes,], family="binomial")
+     summary(mod)$coeff[2,4]
+ })
> names(pval) <- rownames(met[selAPP,])
> sigPval <- which(pval < 0.05/length(pval))
> names(sigPval)
[1] "cg19788250"
```

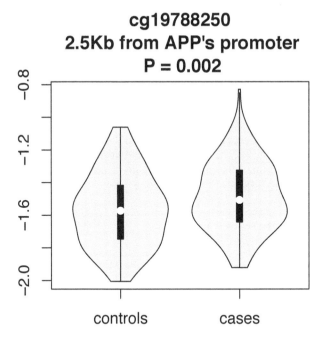

FIGURE 2.6
Methylation differences near APP's promoter between Alzheimer's cases and controls.

We see that one significant association (cg19788250) survives multiple comparisons correction. The location of the CpG is close to the promoter region of the *APP* and in a violin plot, we observe that cases are hypermethylated with respect to controls. While we expect that hypermethylation results in less gene expression, these results are unexpected.

```
>
> #significant CpG info
> cg <- names(sigPval)
> P <- as.character(round(min(pval),3))
>
> #plot
> metapp <- met[cg,]
> lab <- paste0(cg, "\n 2.5Kb from APP's promoter\n P = ", P)
> vioplot(metapp[which(caco==0)], metapp[which(caco==1)],
+          col="gray90", names=c("controls", "cases"))
> title(lab)
```

We can partially explain the unexpected differences in methylation by including relevant covariates in the model, such as tissue, sex and age. If we retrieve these variables and adjust the logistic model by them, we see that the P-value of association between methylation levels and case/control status is no longer significant, after correcting for multiple comparisons. The P-value is incremented to 0.004, when accounting for a significant association with sex ($P = 3 \times 10^{-4}$). Therefore, robust and credible associations need to be consistent with underlying mechanisms, account for multiple comparisons and adjusted for relevant covariates. Adjusting for covariates that may not have been recorded in the study but can be inferred by the *omic* data is the subject of Chapter 6. Once an observation satisfies these criteria then it should be validated in an independent study, further satisfying between observer reproducibility.

```
> #retrieve covariates
> tissue <- pheno$characteristics_ch1
> sex <- pheno$characteristics_ch1.3
> age <- pheno$characteristics_ch1.4
> age <- as.numeric(substr(as.character(age),11,12))
>
> #model of signficant finding adjusted by covariates
> probesig <- selAPP[sigPval]
> mod <- glm(caco ~ met[probesig,]+tissue+sex+age, family="binomial")
> round(summary(mod)$coefficients,4)
```

	Estimate	Std. Error	z value	Pr(>|z|)
(Intercept)	2.0472	1.0932	1.8728	0.0611
met[probesig,]	1.6169	0.5744	2.8148	0.0049
tissuetissue: superior temporal gyrus	0.0205	0.2482	0.0826	0.9342
sexgender: male	-0.9186	0.2557	-3.5922	0.0003
age	0.0101	0.0077	1.3077	0.1910

2.5   Case 3: GTEx

Following our example in Alzheimer's disease, we can then ask which is the expression distribution of *APP* across relevant brain tissues. Bioconductor's package *recount* has made available the transcriptomic data of numerous studies based on RNA-seq [22]. The aim of the package is to reproduce the expression profiles of reported findings, for processing and analysis. In particular, the RNA-seq data from the GTEx project can be easily accessed and analyzed. Expression data for genes across 52 tissues is typically observed in the genome browser of the University of California Santa Cruz (UCSC) or in the GTExwebsite. For *APP* (ENSG00000142192.16) the expression profile plot can be accessed in `https://www.gtexportal.org/home/gene/ENSG00000142192.16`. Here we illustrate how to reproduce the expression profile of *APP* in the cerebellum, brain cortex, frontal cortex and hippocampus using *recount*. The library retrieves the GTEx data using the accession number SRP012682, from the CNA Data Bank of Japan (DDBJ), a public repository of nucleotide sequence data.

```
> library("recount")
```

The data can be downloaded into the working directory for future access.

```
> #download and load data
> download_study('SRP012682', outdir = ".")
```

Data is retrieved by *recount* as a binary R binary file **rse_gene.Rdata**.

```
> load("rse_gene.Rdata")
> #scale RNA-seq data
> rse_gene <- scale_counts(rse_gene)
> rse_gene
```

The data object **rse_gene** contains RNA-seq and genomic annotation data that are extracted by the functions **assays** and **rowRanges**, respectively. We use the annotation data to locate the expression data for *APP* and retrieve the count data for the gene in the **assays** output. For analysis, we add 1 to the count data and **log2** transforms it.

```
> #obtain expression data and annotation
> recountCounts <- assays(rse_gene)$counts
> recountMap <- rowRanges(rse_gene)
>
> #locate data for APP
> indGene <- which(unlist(recountMap$symbol)=="APP")
> selAPP <- grep(names(indGene),rownames(recountCounts))
> recountCounts2APP <- log2(recountCounts[selAPP,]+1)
```

Expression data is then available for all GTEx tissues. We are interested in observing the transcription distribution of the gene in four brain tissues

```
> #select data for four different brain tissues
> gtexPd <- colData(rse_gene)
>
> mask1 <- which(gtexPd$smtsd=="Brain - Cerebellum")
> recountCrb <- recountCounts2APP[mask1]
>
> mask2 <- which(gtexPd$smtsd=="Brain - Cortex")
> recountCtx <- recountCounts2APP[mask2]
>
> mask3 <- which(gtexPd$smtsd=="Brain - Frontal Cortex (BA9)")
> recountFctx <- recountCounts2APP[mask3]
>
> mask4 <- which(gtexPd$smtsd=="Brain - Hippocampus")
> recountHipp <- recountCounts2APP[mask4]
```

Let us plot their distributions in a violin plot

```
> library(vioplot)
> vioplot(recountCrb, recountCtx, recountFctx, recountHipp,
+          col="gray90",
+          names=c("CRB", "CRT", "FRT", "HIP"))
> title("Scaled count data (RNA-seq) for APP")
```

The distributions are consistent to those published in the GTEx portal. In particular, we see that the frontal cortex is the tissue with the highest expression of *APP*. This brain region is affected in later stages of Alzheimer's disease when significant detriment in cognitive abilities are observed. The accessibility of the data encourages further analyses and tissue-specific hypothesis to be tested.

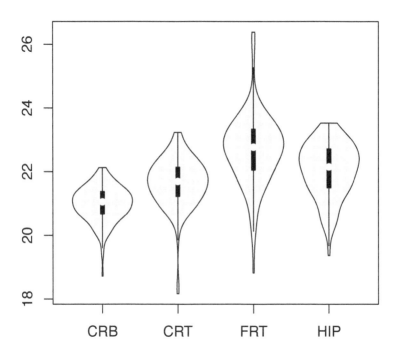

FIGURE 2.7
Expression of APP from RNA-seq count data across four brain tissues (CRB: Cerebellum, CRT: Cortex, FRT: Frontal, HIP: Hippocampus) as measured in the GTEx project.

2.6   Case 4: TCGA

The cancer genome atlas (TCGA) can be fully accessed and analyzed with R/Bioconductor. A wealth of clinical, genomic, transcriptomic and methylomic data, amongst others, are accessible for novel inferences, integration, replication and validation of previous results. We illustrate how to reproduce a reported analysis of a TCGA study in breast cancer. In 2012, the TCGA consortia published a comprehensive analysis of multi *omic* data aimed at characterizing the four main subtypes of breast cancer [101]. Full data for the entire study is available in https://tcga-data.nci.nih.gov/docs/publications/brca_2012/. Data is also available through different Bioconductor's packages, one of which is *RTCGA*. The different *omic* data can be downloaded locally and processed within R, which allows us to reproduce one of the observations reported in for the TCGA breast cancer study. In that study, it is reported that *XBP1* and *ESR1* are genes with the highest differential expression between tumorous and healthy tissues. Interestingly, *XBP1* is important for the immune system and has been shown to interact with the estrogen receptor alpha encoded by *ESR1*.

We retrieve TCGA clinical and RNA-seq data using *RTCGA*. We locate the count data for both genes in tumors and healthy tissues and `log2` transforms them

```
> library(RTCGA.rnaseq)
> library(Biobase)
>
> BRCA.rnaseq_ExpressionSet <- convertTCGA(BRCA.rnaseq)
> genenms <- rownames(BRCA.rnaseq_ExpressionSet)
>
> #locate the full gene names
> genenms[grep("XBP1",genenms)]
[1] "STXBP1|6812" "XBP1|7494"
> genenms[grep("ESR1",genenms)]
[1] "ESR1|2099"
>
> #get expression data
> expr <- expressionsTCGA(BRCA.rnaseq,
+                         extract.cols = c("XBP1|7494","ESR1|2099"))
>
> #get ID data
> barcode <- expr$bcr_patient_barcode
>
> #locate tumor data encoded with 01 string
> cancertissue <- substr(barcode, 14, 15)=="01"
>
> #locate healthy tissue data encoded with 11 string
>
> controltissue <- substr(barcode, 14, 15)=="11"
>
```

```
> #extract ID string relative to subjects
> Subidscancer<-substr(barcode[cancertissue],1,12)
> Subidscontrol<-substr(barcode[controltissue],1,12)
>
> #find which subjects have both cancer and healthy tissue data
> commonids<-intersect(Subidscancer, Subidscontrol)
> selectids<-unlist(sapply(commonids, grep, barcode))
>
> #get expression data for subjects and genes
> expr<-expr[selectids,]
>
> #transfrom
> exprLogXBP1<-log2(expr$`XBP1|7494`+1)
> exprLogESR1<-log2(expr$`ESR1|2099`+1)
```

We first observe that the transcription data between individuals correlate between genes.

```
> plot(exprLogXBP1, exprLogESR1, xlab="log-expression XBP1",
+      ylab="log-expression XBP1", pch=21, bg="gray90")
```

As correlations can be confounded by experimental conditions, we would need to normalize and account of batch effects. While these issues will be cover in Chapter 6, here we can initially confirm the expected co-expression between *XBP1* and *ESR1*. We then would like to see if there are significant gene expression differences between tumors and healthy tissues. We fit a logistic regression between the tumor status and gene expression. The *P*-values of association are obtained for *XBP1* and *ESR1*, respectively. We retrieve data for individuals with tumor and healthy tissue data and create the status outcome variable. Label coding of variables should be consulted in the TCGA documentation (https://gdc.cancer.gov/resources-tcga-users/tcga-code-tables/sample-type-codes).

```
> #outcome variable
> selbarcode <- barcode[selectids]
> status <- rep(NA, length(selectids))
>
> #locate tumor 01 and healty 11 data in selected subjects
> selcancertissue <- substr(selbarcode, 14, 15)=="01"
> selcontroltissue <- substr(selbarcode, 14, 15)=="11"
>
> status[selcancertissue] <- 1
> status[selcontroltissue] <- 0
```

We fit a logistic model for expression differences in *XBP1* and observe a highly significant association

```
> #associations
> mod1 <- glm(status ~ exprLogXBP1, family="binomial")
> res1 <- summary(mod1)$coefficients["exprLogXBP1",]
> res1
    Estimate    Std. Error      z value     Pr(>|z|)
6.509569e-01 1.292194e-01 5.037608e+00 4.713850e-07
```

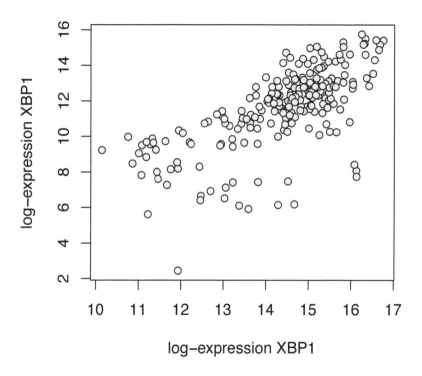

FIGURE 2.8
Correlation between RNA-seq expression levels of XBP1 and ESR1 in tumorous breast tissue.

The association for *ESR1* is, however, less strong and not as significant

```
> mod2 <- glm(status ~ exprLogESR1, family="binomial")
> res2 <- summary(mod2)$coefficients["exprLogESR1",]
> res2
  Estimate Std. Error    z value   Pr(>|z|)
0.16045520 0.06552882 2.44862021 0.01434046
```

A violin plot illustrates the significant differences between healthy and diseased tissues

```
> library(vioplot)
>
> #violin plot
> ##labels
> lab1 <- paste0("XBP1 expression \n P = ",
+                as.character(format(res1["Pr(>|z|)"], sci=TRUE,dig=3)))
>
> lab2 <- paste0("ESR1 expression \n P = ",
+                as.character(round(res2["Pr(>|z|)"],3)))
>
> #remove missing values
> notna<-!is.na(exprLogXBP1) & !is.na(exprLogESR1) & !is.na(status)
>
> par(mfrow=c(1,2), mar=c(5,3,4,1))
> vioplot(exprLogXBP1[which(status==0 & notna)],
+         exprLogXBP1[which(status==1 & notna)],
+         col="gray90", names=c("cont", "tum"))
> title(lab1)
>
> vioplot(exprLogESR1[which(status==0 & notna)],
+         exprLogESR1[which(status==1 & notna)],
+         col="gray90", names=c("cont", "tum"))
> title(lab2)
```

We confirm that, in TCGA data, the expression of *XBP1* and *ESR1* are associated with differences in the cancerous status of the tissue. In particular, expression differences in *XBP1* are stronger than those in *ESR1* and, therefore, further analysis and biological consequences of this observation could be explored. Our aim here was to illustrate how, within R/Bioconductor, we can reproduce previously reported association analysis of *omic* data.

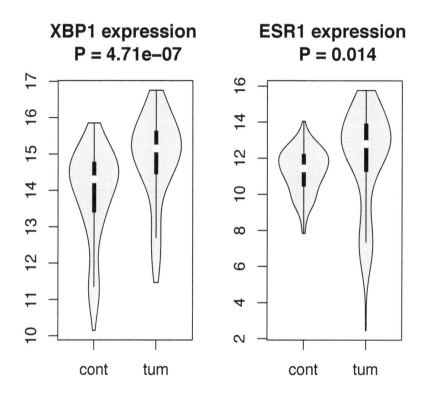

FIGURE 2.9
RNA-seq expression in XBP1 and ESR1 in tumorous breast tissue (tum) and healthy tissue (cont).

2.7   Case 5: NHANES

The National Health and Nutrition Examination Survey (NHANES) is a general population survey from the U.S. (https://wwwn.cdc.gov/nchs/nhanes/). The project offers data on demographic information, physical exam results, laboratory results, such as cholesterol, glucose, or environmental exposures and questionnaire items. It comprises 1,191 variables on 41,474 subjects that can be accessed through the *RNHANES*. The database has exposure assessment to many chemicals such as polycyclic aromatic hydrocarbons (PHA) and health outcomes like cancer.

While natural processes produce PHAs, they are also produced by the incomplete combustion of organic material, making it an important source of pollutant exposure to humans. In particular, PHAs have been linked to cancer through different mechanisms. Here, we illustrate how to access NHANES data and perform an association analysis between cancer and PHAs measurements of exposure. We first install and load the *RNHANES* package

```
> install.packages("RNHANES")
```

that contains the list 1,251 files with the available data that can be downloaded from the NHANES server. A summary of the information can be obtained with **nhanes_data_files**. Specific variables and their description can also be downloaded in R or directly inspected in the website https://wwwn.cdc.gov/nchs/nhanes/search/variablelist.aspx. We search for the files with available information on environmental phenols

```
> library(RNHANES)
> files <- nhanes_data_files()
> #variables <- nhanes_variables()
> nhanes_search(files, "environmental phenols")
       cycle          data_file_description      doc_file
1 2007-2008           Environmental Phenols      EPH_E Doc
2 2003-2004           Environmental Phenols    L24EPH_C Doc
3 2009-2010           Environmental Phenols      EPH_F Doc
4 2005-2006 Environmental Phenols & Parabens    EPH_D Doc
5 2011-2012 Environmental Phenols & Parabens    EPH_G Doc
                      data_file          date_published component
1      EPH_E Data [XPT - 406.9 KB] Updated September 2011 laboratory
2 L24EPH_C Data [XPT - 185.6 KB]       Updated March 2013 laboratory
3      EPH_F Data [XPT - 421.9 KB]            October 2011 laboratory
4       EPH_D Data [XPT - 395 KB]             August 2009 laboratory
5      EPH_G Data [XPT - 347.6 KB]    Updated October 2014 laboratory
  confidential data_file_name file_type data_file_size
1        FALSE          EPH_E       XPT       406.9 KB
2        FALSE        L24EPH_C      XPT       185.6 KB
3        FALSE          EPH_F       XPT       421.9 KB
4        FALSE          EPH_D       XPT         395 KB
5        FALSE          EPH_G       XPT       347.6 KB
```

The file PAH_G contains data of environmental phenols and parabens exposures tested between 2011–2012. Data with demographic information is downloaded with **nhanes_load_data**, indicating the name of the file and the year

```
> nhanes_dat <- nhanes_load_data("PAH_G", "2011-2012", demographics = TRUE)
> rownames(nhanes_dat)<-as.character(nhanes_dat$SEQN)
>
> dim(nhanes_dat)
[1] 2594    74
>
> nhanes_dat[1:10, 1:10]
         SEQN   cycle SDDSRVYR RIDSTATR RIAGENDR RIDAGEYR RIDAGEMN
62168 62168 2011-2012        7        2        1        6       NA
62169 62169 2011-2012        7        2        1       21       NA
62170 62170 2011-2012        7        2        1       15       NA
62171 62171 2011-2012        7        2        1       14       NA
62172 62172 2011-2012        7        2        2       43       NA
62174 62174 2011-2012        7        2        1       80       NA
62178 62178 2011-2012        7        2        1       80       NA
62184 62184 2011-2012        7        2        1       26       NA
62186 62186 2011-2012        7        2        2       17       NA
62189 62189 2011-2012        7        2        2       30       NA
      RIDRETH1 RIDRETH3 RIDEXMON
62168        5        7        1
62169        5        6        1
62170        5        7        1
62171        1        1        1
62172        4        4        1
62174        3        3        2
62178        3        3        2
62184        4        4        1
62186        4        4        1
62189        5        6        1
```

We thus download data for 2594 and 74 variables. A full description of
the variables is accessible at https://wwwn.cdc.gov/nchs/nhanes/Searc
h/default.aspx under the search term PAH_G. We inspect the correlational
structure of PHAs exposures as measured in urine samples. The *survey* and
its addon *jtools* are loaded to perform survey-weighted Pearson correlations
through **nhanes_survey_design**

```
> library(survey)
> library(jtools)
>
> des <- nhanes_survey_design(nhanes_dat)
>
> svycor(~ log(URXP01) + log(URXP02) + log(URXP03) + log(URXP04) +
+         log(URXP05) , design = des, na.rm = TRUE)
            log(URXP01) log(URXP02) log(URXP03) log(URXP04) log(URXP05)
log(URXP01)        1.00        0.58        0.73        0.71        0.63
log(URXP02)        0.58        1.00        0.71        0.74        0.66
log(URXP03)        0.73        0.71        1.00        0.96        0.86
log(URXP04)        0.71        0.74        0.96        1.00        0.88
log(URXP05)        0.63        0.66        0.86        0.88        1.00
```

We can see that the selected PHAs are highly correlated. We then extract a cancer variable MCQ220 from the MCQ file that stores the medical conditions.

```
> medical <- nhanes_load_data("MCQ", "2011-2012", demographics = TRUE)
> rownames(medical) <- as.character(medical$SEQN)
```

We merge the exposure variables, whose names start with URX, with the cancer status (MCQ220), sex (RIAGENDR) and age (RIDAGEYR) within the medical data.

```
> #common names
> nms <- intersect(rownames(medical), rownames(nhanes_dat))
>
> #locate exposures that start with URX
> wh <- grep("URX", colnames(nhanes_dat))
>
> #define PHA matrix exposure
> PAH <- log(nhanes_dat[nms,wh])
> vars <- colnames(PAH)
>
> #select medical variables
> cancer<-medical[nms,c("MCQ220","RIAGENDR","RIDAGEYR")]
>
> #select complete cases
> dat <- data.frame(PAH, cancer)
> selrow <- complete.cases(dat)
> dat<-dat[selrow,]
>
> #Recode cancer=1, control=0
> dat$MCQ220 <- 2-dat$MCQ220
```

We perform a small ExWAS to determine which of these variables are associated with the cancer status

```
> ExWAS<-sapply(vars,  function(pha){
+    fit <- summary(glm(MCQ220 ~ dat[,pha] + RIAGENDR + RIDAGEYR,
+                       data = dat, family = "binomial"))
+
+    fit$coef[2,c(1,2,4)]
+ })
>
> t(ExWAS)
          Estimate Std. Error    Pr(>|z|)
URXP01  0.02621621 0.06065741 0.66559510
URXP02  0.11839709 0.08346592 0.15604245
URXP03  0.16361702 0.07160616 0.02231534
URXP04  0.16389201 0.08032554 0.04131573
URXP05  0.13757055 0.09132980 0.13198887
URXP06  0.18367871 0.10228259 0.07252690
URXP07  0.13042280 0.10174588 0.19989582
URXP10  0.14711180 0.08845188 0.09627549
URXP17  0.08337620 0.08840826 0.34563919
URXP19  0.09863729 0.10655709 0.35461464
URXUCR -0.14306344 0.13852354 0.30170990
```

We observe that URXP03 is significantly associated with cancer status at a nominal level. URXP04 is also associated with cancer due to the high correlation with URXP03, as previously observed. However, URXP04 corresponds to the measurements of 2-hydroxyfluorene in urine which has been shown to be an efficient biomarker for evaluating the exposure to PAHs from smoking [157].

3

Dealing with omic data in Bioconductor

CONTENTS

3.1 Chapter overview

This chapter offers a summary of the main data structures that are implemented in Bioconductor for dealing with genomic, transcriptomic, methylomic and exposomic data. The structures are objects to which methods are applied. *Omic* data is typically composed of three datasets: one containing the actual high-dimensional data of *omic* variables per individuals, annotation data that specifies the characteristics of the variables and phenotypic information that encodes the subject's traits of interest, covariates and sampling characteristics. For instance, transcriptomic data is stored in a *ExpressionSet* object, which is a data structure that contains the transcription values of individuals at each transcription probe, the genomic information for the transcription probes and the phenotypes of the individuals. Specific data is accessed, processed and analyzed with specific functions from diverse packages, conceived as methods acting on the *ExpressionSet* object. The aim of this chapter is then to introduce the specific *omic* objects available in Bioconductor. In the following chapters, we will introduce the packages that have been implemented to process and analyze these objects.

3.2 *snpMatrix*

SNP array data can be stored in different formats. PLINK binary format
(http://www.cog-genomics.org/plink2) is a common and efficient system
to store and analyze genotype data. It was developed to analyze data with
PLINK software [115] but its efficiency in storing data in binary files has made
it one of the standard formats for other software packages, including some in
Bioconductor. PLINK stores SNP genomic data, annotation and phenotype
information in three different files with extensions .bed, .bim and .fam:

- binary BED file: Contains the genomic SNP data, whose values are encoded in two bits (Homozygous normal 00, Heterozygous 10, Homozygous
  variant 11, missing 01).

- text BIM file: Contains SNPs annotations. Each row is a SNP and contains
  six columns: chromosome, SNP name, position in morgans, base-pair coordinates, allele 1 (reference nucleotide), allele 2 (alternative nucleotide).

- text FAM file: Contains the subject's information. Each row is an individual and contains six variables: the Family identifier (ID), Individual
  ID, Paternal ID, Maternal ID, Sex (1=male; 2=female; other=unknown),
  phenotypes. Covariates can be added in additional columns.

PLINK data can be loaded into R with the read.plink function from
snpStats Bioconductor's package. The function requires the full path for the
BED, BIM and FAM files, or only their name when the working directory of
R contains the PLINK files.

```
> library(snpStats)
> ob.plink <- read.plink(bed = "obesity.bed",
+                         bim = "obesity.bim",
+                         fam = "obesity.fam")
```

In case of having the three files with the same name (i.e. obesity.fam,
obesity.bim and obesity.bed) the following simplification may be convenient

```
> snps <- read.plink("obesity")
```

The files used in this example are stored in the *brgedata* package. Therefore,
they can also be loaded by

```
> path <- system.file("extdata", package="brgedata")
> snps <- read.plink(file.path(path, "obesity"))
> names(snps)
[1] "genotypes" "fam"        "map"
```

The `read.plink` function returns a list with the three fields `genotypes`, `fam` and `map` that correspond to the three uploaded files. The `genotypes` field contains the genotype data stored in a *snpMatrix* object (individuals in rows and SNPs in columns).

```
> geno <- snps$genotypes
> geno
A SnpMatrix with  2312 rows and  100000 columns
Row names:  100 ... 998
Col names:  MitoC3993T ... rs28600179
```

Genotypes are encoded as raw variables for storage efficiency. While individual values can be extracted with array syntax, manipulation of data is usually performed by methods that act on the complete object. The `fam` field contains the individual's information in a *data.frame* object:

```
> individuals <- snps$fam
> head(individuals)
     pedigree member father mother sex affected
100    FAM_OB    100     NA     NA   1        1
1001   FAM_OB   1001     NA     NA   1        1
1004   FAM_OB   1004     NA     NA   2        2
1005   FAM_OB   1005     NA     NA   1        2
1006   FAM_OB   1006     NA     NA   2        1
1008   FAM_OB   1008     NA     NA   1        1
```

The `map` field contains the SNPs annotation in a *data.frame*:

```
> annotation <- snps$map
> head(annotation)
            chromosome   snp.name cM position allele.1 allele.2
MitoC3993T          NA MitoC3993T NA     3993        T        C
MitoG4821A          NA MitoG4821A NA     4821        A        G
MitoG6027A          NA MitoG6027A NA     6027        A        G
MitoT6153C          NA MitoT6153C NA     6153        C        T
MitoC7275T          NA MitoC7275T NA     7275        T        C
MitoT9699C          NA MitoT9699C NA     9699        C        T
```

Subsetting SNP data requires at least two different operations. For instance, if we are interested in extracting the SNPs of chromosome 1, we need to select those variants in `annotation` that are located in chromosome 1 and then subset the *snpMatrix* object as a typical R *matrix*

```
> annotationChr1 <- annotation[annotation$chromosome == "1" &
+                     !is.na(annotation$chromosome), ]
> genoChr1 <- geno[, rownames(annotationChr1)]
> genoChr1
A SnpMatrix with  2312 rows and  7721 columns
Row names:  100 ... 998
Col names:  rs12354060 ... rs7527472
```

Subsetting samples follow a similar pattern. Suppose we want to select the genotypes of the control individuals. Case-control status is often encoded in the FAM file and therefore uploaded in the `fam` field. In our example, controls are coded with 1 and cases with 2 in the variable `affected` of `individuals`. Therefore, the genotypes of the control samples are extracted by

```
> individualsCtrl <- individuals[individuals$affected == 1, ]
> genoCtrl <- geno[rownames(individualsCtrl), ]
> genoCtrl
A SnpMatrix with  1587 rows and  100000 columns
Row names:  100 ... 998
Col names:  MitoC3993T ... rs28600179
```

3.3 *ExpressionSet*

ExpressionSet was one of the first implementations of Bioconductor to manage *omic* experiments. Although its use is discouraged in Bioconductor's guidelines for the development of current and future packages, most publicly available data is available in this structure while future packages are still required to be able to upload and operate with it. The GEO repository contains thousands of transcriptomic experiments that are available in *ExpressionSet* format. Take our Case 2 of Chapter 2, where we download the data of the experiment with accession number GSE63061 from the GEO website

```
> library(GEOquery)
> gsm.expr <- getGEO("GSE63061", destdir = ".")[[1]]
> gsm.expr <- gsm.expr
```

`gsm.expr` is an object of class *ExpressionSet* that has three main slots. Transcriptomic data is stored in the `assayData` slot, phenotypes are in `phenoData` and probe annotation in `featuredData`. There are three other slots `protocolData`, `experimentData` and `annotation` that specify equipment-generated information about protocols, resulting publications and the platform on which the samples were assayed. Methods are implemented to extract the data from each slot of the object. For instance `exprs` extracts the trancriptomic data in a matrix where subjects are columns and probes are rows

```
> expr <- exprs(gsm.expr)
> dim(expr)
[1] 32049    388
> expr[1:5,1:5]
            GSM1539409 GSM1539410 GSM1539411 GSM1539412 GSM1539413
ILMN_1343291  12.552807  12.711459  13.088393  12.643831  13.098389
ILMN_1343295  10.101556   9.776015   9.594397  10.126782  10.223301
ILMN_1651209   6.084671   6.255012   6.160485   6.109219   6.069960
ILMN_1651210   6.068805   6.016468   6.024322   6.016118   6.056163
ILMN_1651221   6.121060   6.173167   6.039552   6.111306   6.089542
```

phenoData retrieves the subjects' phenotypes in an *AnnotatedDataFrame* object which is converted to *data.frame* by the function **pData**

```
> #get phenotype data
> pheno <- phenoData(gsm.expr)
> pheno
An object of class 'AnnotatedDataFrame'
  sampleNames: GSM1539409 GSM1539410 ... GSM1539796 (388 total)
  varLabels: title geo_accession ... tissue:ch1 (40 total)
  varMetadata: labelDescription
> phenoDataFrame <- pData(gsm.expr)
> phenoDataFrame[1:5,1:4]
                       title geo_accession                status
GSM1539409 7196843065_F        GSM1539409 Public on Aug 05 2015
GSM1539410 7196843076_G        GSM1539410 Public on Aug 05 2015
GSM1539411 7196843068_B        GSM1539411 Public on Aug 05 2015
GSM1539412 7196843063_B        GSM1539412 Public on Aug 05 2015
GSM1539413 7196843065_L        GSM1539413 Public on Aug 05 2015
           submission_date
GSM1539409       Nov 06 2014
GSM1539410       Nov 06 2014
GSM1539411       Nov 06 2014
GSM1539412       Nov 06 2014
GSM1539413       Nov 06 2014
>
> #Alzheimer's case control variable
> summary(phenoDataFrame$characteristics_ch1)
             status: AD status: borderline MCI            status: CTL
                    139                      3                    134
      status: CTL to AD          status: MCI     status: MCI to CTL
                      1                    109                      1
          status: OTHER
                      1
```

Finally the **fData** function gets the probes' annotation in a *data.frame*

```
> probes <- fData(gsm.expr)
> probes[1:5, 1:5]
                         ID     Species Source  Search_Key   Transcript
ILMN_1343291 ILMN_1343291 Homo sapiens RefSeq NM_001402.4   ILMN_5311
ILMN_1343295 ILMN_1343295 Homo sapiens RefSeq              ILMN_27206
ILMN_1651209 ILMN_1651209 Homo sapiens RefSeq NM_182838.1   ILMN_8692
ILMN_1651210 ILMN_1651210 Homo sapiens RefSeq XM_941691.1 ILMN_138115
ILMN_1651221 ILMN_1651221 Homo sapiens RefSeq XM_926225.1  ILMN_33528
```

3.4   *SummarizedExperiment*

The *SummarizedExperiment* class is a comprehensive data structure that can be used to store expression and methylation data from microarrays or read counts from RNA-seq experiments, among others. A *SummarizedExperiment* object contains slots for one or more *omic* datasets, feature annotation (e.g. genes, transcripts, SNPs, CpGs), individual phenotypes and experimental details, such as laboratory and experimental protocols. In a *SummarizedExperiment*, the rows of *omic* data are features and columns are subjects.

Information is coordinated across the object's slots. For instance, subsetting samples in the assay matrix automatically subsets them in the phenotype metadata. A *SummarizedExperiment* object is easily manipulated and constitutes the input and output of many of Bioconductor's methods.

Data is retrieved from a *SummarizedExperiment* by using specific methods or accessors. We illustrate the functions with **brge_methy** which includes real methylation data and is available from the *brgedata* package. The data is made available by loading of the package

```
> library(brgedata)
> brge_methy
class: GenomicRatioSet
dim: 392277 20
metadata(0):
assays(1): Beta
rownames(392277): cg13869341 cg24669183 ... cg26251715 cg25640065
rowData names(14): Forward_Sequence SourceSeq ...
  Regulatory_Feature_G DHS
colnames(20): x0017 x0043 ... x0077 x0079
colData names(9): age sex ... Mono Neu
Annotation
  array: IlluminaHumanMethylation450k
  annotation: ilmn12.hg19
Preprocessing
  Method: NA
  minfi version: NA
  Manifest version: NA
> extends("GenomicRatioSet")
[1] "GenomicRatioSet"          "RangedSummarizedExperiment"
[3] "SummarizedExperiment"     "Vector"
[5] "Annotated"
```

The function **extends** shows that the data has been encoded in an object of *GenomicRatioSet* class, which is an extension of the more primitive classes *RangedSummarizedExperiment* and *SummarizedExperiment*. **brge_methy** illustrates a typical object within Bioconductor's framework, as it is a structure that inherits different types of classes in an established hierarchy. For each class, there are specific methods which are properly inherited across classes. For instance, in our example, *SummarizedExperiment* is **brge_methy**'s most primitive *omic* class and, therefore, all the methods of *SummarizedExperiment* apply to it. In particular, the methylation data that is stored in the object can be extracted with the function **assay**

```
> betas <- assay(brge_methy)
> betas[1:5, 1:4]
                x0017      x0043     x0036      x0041
cg13869341 0.90331003 0.90082827 0.8543610 0.84231205
```

```
cg24669183 0.86349463 0.84009629 0.8741255 0.85122181
cg15560884 0.65986032 0.69634338 0.6931083 0.69752294
cg01014490 0.01448404 0.01509478 0.0163614 0.01322362
cg17505339 0.92176287 0.92060149 0.9101060 0.93249167
```

The assay slot of a *SummarizedExperiment* object can contain any type of data (i.e. numeric, character, factor...), structure or large on-disk representations, such as a HDF5Array. Feature annotation data is accessed with the function `rowData`:

```
> rowData(brge_methy)[,2:5]
DataFrame with 392277 rows and 4 columns
                                                  SourceSeq Random_Loci
                                                <character> <character>
1       CCGGTGGCTGGCCACTCTGCTAGAGTCCATCCGCCAAGCTGGGGGCATCG
2       TCACCGCCTTGACAGCTTTGCAGAGTGCTGCTCAGGTATTCTGCAAGACG
3       CGCGTAAACAAGGGAAGCTGAGTAATTGTATGTTCAAATACTTGCAAAAC
4       TCAGAACTCGCGGTGGGGGCTGCTGGTTCTTCCAGGAGCGCGCATGAGCG
5       AAACAAACAAAGATATCAAGCCACAGATTCAAAGTGCTATAAACTCCACG
...                                                    ...         ...
392273  CGGCGGCTTTCCACGCTGCGGCTTGGAGTGGTCCTTGTTTAGATTCCTTT
392274  CGCAGGTATGGTGTACATTAAGCAGGCAGGGTCAATCAGGGATGGTCTAT
392275  CAAAATGAATGAAATTTACAAACCCAGCAGCCAATAGATCTCCAGAGTCG
392276  CGCTGGTTGTCCAGGCTGGAATGCAATGGTGCAATGTCCGCTCACTCCAA
392277  CGGGGCAGGGGTCCGTAACTGCAGCCCTCCATGCCTGAGCCCCCCACCCC
            Methyl27_Loci         genes
              <character>   <character>
1                                 WASH5P
2
3
4
5
...                 ...           ...
392273
392274
392275                      NLGN4Y;NLGN4Y
392276                            TTTY14
392277
```

which returns a `data.frame` object. In our example, it contains the sequences and the genes associated with the CpG probes, among other information.

3.5 *GRanges*

The Bioconductor's package *GenomicRanges* aims to represent and manipulate the genomic annotation of molecular *omic* data under a reference genome. It contains functions to select specific regions and perform operations with them [76]. Objects of *GRanges* class are important to annotate and manipulate genomic, transcriptomic and methylomic data. In particular, they are used in conjunction with *SummarizedExperiment*, within the *RangedSumma-rizedExperiment* class that is explained in the following section.

Annotation data refers to the characteristics of the variables in the high-dimensional data set. In particular for *omic* data relative to DNA structure and function, each variable may be given a location in a reference genome. While not two genomes are identical, the construction of a reference genome allows the mapping of specific characteristics of individual genomes to a common ground where they can be compared. The reference genome defines a coordinate system: chromosome id and position along the chromosome. For instance, a position such as chr10:4567-5671 would represent the 4567th to the 5671st base pair on the reference's chromosome 10.

The main functionalities implemented *GenomicRanges* are methods on *GRanges* objects. Objects are created by the function GRanges, minimum requirements are the genomic positions given by the chromosome (seqnames) and base pair coordinates (ranges). Other metadata (e.g. variables) can be added to provide further information about each segment.

We illustrate *GenomicRanges* creating 8 segments on either chr1 or chr2, each with defined start and end points. We add strand information, passed through the argument strand, to indicate the direction of each sequence. We also add a hypothetical variable disease that indicates whether asthma or obesity have been associated with each interval

```
> library(GenomicRanges)
> gr <- GRanges(seqnames=c(rep("chr1", 4), rep("chr2", 4)),
+               ranges = IRanges(start = c(1000, 1800, 5300, 7900,
+                                          1300, 2100, 3400, 6700),
+                                end =c(2200, 3900, 5400, 8100,
+                                       2600, 3300, 4460, 6850)),
+               strand = rep(c("+", "-"), 4),
+               disease = c(rep("Asthma",4), rep("Obesity",4)))
> gr
GRanges object with 8 ranges and 1 metadata column:
      seqnames    ranges strand |      disease
         <Rle> <IRanges>  <Rle> |  <character>
  [1]     chr1 1000-2200      + |       Asthma
  [2]     chr1 1800-3900      - |       Asthma
  [3]     chr1 5300-5400      + |       Asthma
  [4]     chr1 7900-8100      - |       Asthma
  [5]     chr2 1300-2600      + |      Obesity
  [6]     chr2 2100-3300      - |      Obesity
  [7]     chr2 3400-4460      + |      Obesity
  [8]     chr2 6700-6850      - |      Obesity
  -------
  seqinfo: 2 sequences from an unspecified genome; no seqlengths
```

gr is our constructed object of class *GRanges*. Within Bioconductor there are numerous databases that can be used to create *GRanges* objects based on the reference genomes of many species, see *biomaRt* in Sections 3.9, 4.3.1 and *Gviz* in Section 4.6. *GRanges* objects can be large and there are several functions to access data in different ways. For instance, the *GRanges* object responds to the usual array and subset extraction given by squared parentheses

```
> gr[1]
GRanges object with 1 range and 1 metadata column:
      seqnames    ranges strand |    disease
         <Rle> <IRanges>  <Rle> | <character>
  [1]      chr1 1000-2200      + |     Asthma
  -------
  seqinfo: 2 sequences from an unspecified genome; no seqlengths
```

However, there are also specific functions to access and modify information. For instance, **seqnames** extract the chromosomes defined in our examples, whose first element can be redefined accordingly:

```
> seqnames(gr)
factor-Rle of length 8 with 2 runs
  Lengths:    4    4
  Values :  chr1 chr2
Levels(2):  chr1 chr2
> seqnames(gr)[1] <- "chr2"
> gr
GRanges object with 8 ranges and 1 metadata column:
      seqnames    ranges strand |    disease
         <Rle> <IRanges>  <Rle> | <character>
  [1]      chr2 1000-2200      + |     Asthma
  [2]      chr1 1800-3900      - |     Asthma
  [3]      chr1 5300-5400      + |     Asthma
  [4]      chr1 7900-8100      - |     Asthma
  [5]      chr2 1300-2600      + |    Obesity
  [6]      chr2 2100-3300      - |    Obesity
  [7]      chr2 3400-4460      + |    Obesity
  [8]      chr2 6700-6850      - |    Obesity
  -------
  seqinfo: 2 sequences from an unspecified genome; no seqlengths
```

Additional information can be added to the current object as a new field of a **list**

```
> gr$gene_id <- paste0("Gene", 1:8)
> gr
GRanges object with 8 ranges and 2 metadata columns:
      seqnames    ranges strand |    disease     gene_id
         <Rle> <IRanges>  <Rle> | <character> <character>
  [1]      chr2 1000-2200      + |     Asthma       Gene1
  [2]      chr1 1800-3900      - |     Asthma       Gene2
  [3]      chr1 5300-5400      + |     Asthma       Gene3
  [4]      chr1 7900-8100      - |     Asthma       Gene4
  [5]      chr2 1300-2600      + |    Obesity       Gene5
  [6]      chr2 2100-3300      - |    Obesity       Gene6
  [7]      chr2 3400-4460      + |    Obesity       Gene7
  [8]      chr2 6700-6850      - |    Obesity       Gene8
  -------
  seqinfo: 2 sequences from an unspecified genome; no seqlengths
```

GenomicRanges provides different methods to perform arithmetic with the ranges, see *?GRanges* for a full list. For instance, with `shift` an interval is moved a given base-pair distance and with `flank` the interval is stretched

```
> #shift: move all intervals 10 base pair towards the end
> shift(gr, 10)
GRanges object with 8 ranges and 2 metadata columns:
      seqnames    ranges strand |      disease     gene_id
         <Rle> <IRanges>  <Rle> |  <character> <character>
  [1]     chr2 1010-2210      + |       Asthma       Gene1
  [2]     chr1 1810-3910      - |       Asthma       Gene2
  [3]     chr1 5310-5410      + |       Asthma       Gene3
  [4]     chr1 7910-8110      - |       Asthma       Gene4
  [5]     chr2 1310-2610      + |      Obesity       Gene5
  [6]     chr2 2110-3310      - |      Obesity       Gene6
  [7]     chr2 3410-4470      + |      Obesity       Gene7
  [8]     chr2 6710-6860      - |      Obesity       Gene8
  -------
  seqinfo: 2 sequences from an unspecified genome; no seqlengths
>
> #shift: move each intervals individually
> shift(gr, seq(10,100, length=8))
GRanges object with 8 ranges and 2 metadata columns:
      seqnames    ranges strand |      disease     gene_id
         <Rle> <IRanges>  <Rle> |  <character> <character>
  [1]     chr2 1010-2210      + |       Asthma       Gene1
  [2]     chr1 1822-3922      - |       Asthma       Gene2
  [3]     chr1 5335-5435      + |       Asthma       Gene3
  [4]     chr1 7948-8148      - |       Asthma       Gene4
  [5]     chr2 1361-2661      + |      Obesity       Gene5
  [6]     chr2 2174-3374      - |      Obesity       Gene6
  [7]     chr2 3487-4547      + |      Obesity       Gene7
  [8]     chr2 6800-6950      - |      Obesity       Gene8
  -------
  seqinfo: 2 sequences from an unspecified genome; no seqlengths
>
> #flank:  recover regions next to the input set.
> #        For a 50 base stretch upstream (negative value for
> #        downstream)
> flank(gr, 50)
GRanges object with 8 ranges and 2 metadata columns:
      seqnames    ranges strand |      disease     gene_id
         <Rle> <IRanges>  <Rle> |  <character> <character>
  [1]     chr2   950-999      + |       Asthma       Gene1
  [2]     chr1 3901-3950      - |       Asthma       Gene2
  [3]     chr1 5250-5299      + |       Asthma       Gene3
  [4]     chr1 8101-8150      - |       Asthma       Gene4
  [5]     chr2 1250-1299      + |      Obesity       Gene5
  [6]     chr2 3301-3350      - |      Obesity       Gene6
  [7]     chr2 3350-3399      + |      Obesity       Gene7
  [8]     chr2 6851-6900      - |      Obesity       Gene8
  -------
  seqinfo: 2 sequences from an unspecified genome; no seqlengths
```

GenomicRanges also includes methods for aggregating and summarizing *GRanges* objects. `reduce`, `disjoint` and `coverage` methods are most useful.

disjoin, for instance, reduces the intervals into the smallest set of unique, non-overlapping pieces that make up the original object. It is strand-specific by default, but this can be avoided with `ignore.strand=TRUE`

```
> disjoin(gr, ignore.strand=TRUE)
GRanges object with 10 ranges and 0 metadata columns:
       seqnames      ranges strand
          <Rle>   <IRanges>  <Rle>
   [1]     chr1 1800-3900      *
   [2]     chr1 5300-5400      *
   [3]     chr1 7900-8100      *
   [4]     chr2 1000-1299      *
   [5]     chr2 1300-2099      *
   [6]     chr2 2100-2200      *
   [7]     chr2 2201-2600      *
   [8]     chr2 2601-3300      *
   [9]     chr2 3400-4460      *
  [10]     chr2 6700-6850      *
  -------
  seqinfo: 2 sequences from an unspecified genome; no seqlengths
```

reduce creates the smallest range set of unique, non-overlapping intervals. Strand information is also taken into account by default and can also be turned off

```
> reduce(gr, ignore.strand=TRUE)
GRanges object with 6 ranges and 0 metadata columns:
      seqnames      ranges strand
         <Rle>   <IRanges>  <Rle>
  [1]     chr1 1800-3900      *
  [2]     chr1 5300-5400      *
  [3]     chr1 7900-8100      *
  [4]     chr2 1000-3300      *
  [5]     chr2 3400-4460      *
  [6]     chr2 6700-6850      *
  -------
  seqinfo: 2 sequences from an unspecified genome; no seqlengths
```

coverage summarizes the times each base is covered by an interval

```
> coverage(gr)
RleList of length 2
$chr1
integer-Rle of length 8100 with 6 runs
  Lengths: 1799 2101 1399  101 2499  201
  Values :    0    1    0    1    0    1

$chr2
integer-Rle of length 6850 with 10 runs
  Lengths:  999  300  800  101  400  700   99 1061 2239  151
  Values :    0    1    2    3    2    1    0    1    0    1
```

It is also possible to perform operations between two different GRanges objects. For instance, one may be interested in knowing the intervals that overlap with a targeted region:

```
> target <- GRanges(seqnames="chr1",
+                   range=IRanges(start=1200, 4000))
> target
GRanges object with 1 range and 0 metadata columns:
      seqnames    ranges strand
         <Rle> <IRanges>  <Rle>
  [1]     chr1 1200-4000      *
  -------
  seqinfo: 1 sequence from an unspecified genome; no seqlengths
> gr.ov <- findOverlaps(target, gr)
> gr.ov
Hits object with 1 hit and 0 metadata columns:
      queryHits subjectHits
      <integer>   <integer>
  [1]         1           2
  -------
  queryLength: 1 / subjectLength: 8
```

To recover the overlapping intervals between **gr** and **target** we can run

```
> gr[subjectHits(gr.ov)]
GRanges object with 1 range and 2 metadata columns:
      seqnames    ranges strand |     disease     gene_id
         <Rle> <IRanges>  <Rle> | <character> <character>
  [1]     chr1 1800-3900      - |      Asthma       Gene2
  -------
  seqinfo: 2 sequences from an unspecified genome; no seqlengths
```

The Table 3.1 shows the common operations of **GRanges**

TABLE 3.1

Common operations on *IRanges*, *GRanges* and *GRangesList*. Table obtained from `https://bioconductor.org/help/course-materials/2014/SeattleFeb2014/`.

Category	Function	Description
Accessors	`start, end, width`	Get or set the starts, ends and widths
	`names`	Get or set the names
	`mcols, metadata`	Get or set metadata on elements or object
	`length`	Number of ranges in the vector
	`range`	Range formed from min start and max end
Ordering	`<, <=, >, >=, ==, !=`	Compare ranges, ordering by start then width
	`sort, order, rank`	Sort by the ordering
	`duplicated`	Find ranges with multiple instances
	`unique`	Find unique instances, removing duplicates
Arithmetic	`r + x, r - x, r * x`	Shrink or expand ranges `r` by number `x`
	`shift`	Move the ranges by specified amount
	`resize`	Change width, anchoring on start, end or mid
	`distance`	Separation between ranges (closest endpoints)
	`restrict`	Clamp ranges to within some start and end
	`flank`	Generate adjacent regions on start or end
Set operations	`reduce`	Merge overlapping and adjacent ranges
	`intersect, union, setdiff`	Set operations on reduced ranges
	`pintersect, punion, psetdiff`	Parallel set operations, on each `x[i], y[i]`
	`gaps, pgap`	Find regions not covered by reduced ranges
	`disjoin`	Ranges formed from union of endpoints
Overlaps	`findOverlaps`	Find all overlaps for each `x` in `y`
	`countOverlaps`	Count overlaps of each `x` range in `y`
	`nearest`	Find nearest neighbors (closest endpoints)
	`precede, follow`	Find nearest `y` that `x` precedes or follows
	`x %in% y`	Find ranges in `x` that overlap range in `y`
Coverage	`coverage`	Count ranges covering each position
Extraction	`r[i]`	Get or set by logical or numeric index
	`r[[i]]`	Get integer sequence from `start[i]` to `end[i]`
	`subsetByOverlaps`	Subset `x` for those that overlap in `y`
	`head, tail, rev, rep`	Conventional R semantics
Split, combine	`split`	Split ranges by a factor into a *RangesList*
	`c`	Concatenate two or more range objects

3.6   RangedSummarizedExperiment

SummarizedExperiment is extended to *RangedSummarizedExperiment*, a child
class that contains the annotation data of the features in a `GenomicRanges`
object. In our methylomic example, the second most primitive class of
`brge_methy` with *omic* functionality, after *SummarizedExperiment*, is *Ranged-*
SummarizedExperiment. Annotation data, with variable names given by

```
> names(rowData(brge_methy))
 [1] "Forward_Sequence"       "SourceSeq"
 [3] "Random_Loci"            "Methyl27_Loci"
 [5] "genes"                  "UCSC_RefGene_Accession"
 [7] "group"                  "Phantom"
 [9] "DMR"                    "Enhancer"
[11] "HMM_Island"             "Regulatory_Feature_Name"
[13] "Regulatory_Feature_G" "DHS"
```

can be obtained in a *GRanges* object, for a given variable. For instance, metadata of CpG genomic annotation and neighbouring genes is obtained using
array syntax

```
> rowRanges(brge_methy)[, "genes"]
GRanges object with 392277 ranges and 1 metadata column:
              seqnames      ranges strand |               genes
                 <Rle>   <IRanges>  <Rle> |         <character>
  cg13869341      chr1       15865      * |               WASH5P
  cg24669183      chr1      534242      * |
  cg15560884      chr1      710097      * |
  cg01014490      chr1      714177      * |
  cg17505339      chr1      720865      * |
         ...       ...         ...    ... .                 ...
  cg04964672      chrY     7428198      * |
  cg01086462      chrY     7429349      * |
  cg02233183      chrY    16634382      * | NLGN4Y;NLGN4Y
  cg26251715      chrY    21236229      * |               TTTY14
  cg25640065      chrY    23569324      * |
  -------
  seqinfo: 24 sequences from an unspecified genome; no seqlengths
```

Subject data can be accessed entirely in a single *data.frame* or a variable
at the time. The entire subject (phenotype) information is retrieved with the
function `colData`:

```
> colData(brge_methy)
DataFrame with 20 rows and 9 columns
            age      sex                    NK               Bcell
        <numeric> <factor>             <numeric>           <numeric>
x0017         4   Female -5.81386004885394e-19   0.173507848676514
x0043         4   Female  0.00164826459258012    0.182017163135389
x0036         4     Male  0.0113186314286605     0.169017297114852
x0041         4   Female  0.00850822150291592    0.0697157716842748
x0032         4     Male                     0    0.113977975025216
```

```
...          ...        ...                ...                   ...
x0018         4      Female     0.0170284419499822 0.0781697956157343
x0057         4      Female                      0 0.0797774108319114
x0061         4      Female                      0  0.164026603367578
x0077         4      Female                      0   0.11227306633587
x0079         4      Female     0.0120147521727839  0.091365002951843
                               CD4T                 CD8T              Eos
                          <numeric>            <numeric>        <numeric>
x0017     0.204068686112657    0.100974056859719                0
x0043     0.161712715578477    0.128772235710367                0
x0036     0.155463683913806     0.12774174901185                0
x0041 0.00732789124320035   0.0321860552286678                  0
x0032     0.222303988949171   0.0216089881163733 2.60503891663895e-18
...                 ...                  ...                  ...
x0018     0.112735084403202   0.0679681631510725   0.00837769947260005
x0057     0.111071701794334 0.00910488689520549 4.61887343424789e-18
x0061     0.224202761557123    0.131252122776747   0.0085107393064941
x0077     0.168056145550346    0.078405927598852    0.0613396950863682
x0079     0.205830330328636    0.114753887947108                0
                               Mono                 Neu
                          <numeric>            <numeric>
x0017 0.0385653506040031    0.490936252606603
x0043  0.049954190999866    0.491822214741198
x0036  0.102710482176672    0.459631855532962
x0041 0.0718279616580959    0.807749226994262
x0032 0.0567246148936481    0.575614453750378
...                 ...                  ...
x0018 0.0579534708330382    0.658600075099212
x0057  0.101526022615966    0.686010320431362
x0061 0.0382646576955094    0.429574723586544
x0077 0.0583411264640133    0.515284487718098
x0079 0.0750535276475847    0.513596840714489
```

The *list* symbol $ can be used, for instance, to obtain the sex of the individuals

```
> brge_methy$sex
 [1] Female Female Male   Female Male    Male   Male   Male   Male   Male
[11] Female Female Female Male   Male    Female Female Female Female Female
Levels: Female Male
```

Subsetting the entire structure is also possible following the usual array syntax. For example, we can select only males from the **brge_methy** dataset

```
> brge_methy[, brge_methy$sex == "male"]
class: GenomicRatioSet
dim: 392277 0
metadata(0):
assays(1): Beta
rownames(392277): cg13869341 cg24669183 ... cg26251715 cg25640065
rowData names(14): Forward_Sequence SourceSeq ...
  Regulatory_Feature_G DHS
colnames: NULL
colData names(9): age sex ... Mono Neu
Annotation
```

```
array: IlluminaHumanMethylation450k
annotation: ilmn12.hg19
Preprocessing
Method: NA
minfi version: NA
Manifest version: NA
```

The `metadata` function retrieves experimental data

```
> metadata(brge_methy)
list()
```

which in our case is empty.

3.7   *ExposomeSet*

The *rexposome* package provides a structure to store exposomic data in Bioconductor's framework. *ExposomeSet* is the class that encapsulates exposomic data with three slots that store the exposomic matrix, the description of the exposures and the phenotypic information. An *ExposomeSet* object can be created from data stored in three text files that correspond to each dataset.

We illustrate how to construct an *ExposomeSet* object from data stored in csv files, which are distributed with the *rexposome* package,

```
> library(rexposome)
> path <- system.file("extdata", package="rexposome")
> description <- file.path(path, "description.csv")
> exposures <- file.path(path, "exposures.csv")
> phenotype <- file.path(path, "phenotypes.csv")
```

The *readExposome* function loads these data into an *ExposomeSet* object. By default it reads csv format that can be changed with the argument `sep`. Missing strings are identified with the argument `na.strings`. The arguments `exposures.samCol` and `phenotype.samCol` indicate the variables (columns) in the exposure and phenotype files with the subjects' IDs. The arguments `description.expCol` and `description.famCol` indicate the columns containing the exposures names and the exposures families in the description file

```
> expo <- readExposome(exposures = exposures,
+                      description = description,
+                      phenotype = phenotype,
+                      exposures.samCol = "idnum",
+                      description.expCol = "Exposure",
+                      description.famCol = "Family",
+                      phenotype.samCol = "idnum"
+ )
```

  expo is an object of class *ExposomeSet*

```
> expo
Object of class 'ExposomeSet' (storageMode: environment)
 . exposures description:
   . categorical:  4
   . continuous:  84
 . exposures transformation:
   . categorical: 0
   . transformed: 0
   . standardized: 0
   . imputed: 0
 . assayData: 88 exposures 109 individuals
   . element names: exp
   . exposures: AbsPM25, ..., X7OHMMeOP
   . individuals: id001, ..., id108
 . phenoData: 109 individuals 9 phenotypes
   . individuals: id001, ..., id108
   . phenotypes: whistling_chest, ..., cbmi
 . featureData: 88 exposures 7 explanations
   . exposures: AbsPM25, ..., X7OHMMeOP
   . descriptions: Family, ..., .imp
experimentData: use 'experimentData(object)'
Annotation:
```

with all the exposomic, exposure annotation and phenotype data. We see
that there are 84 continuous and 4 categorical exposures. The `assayData`,
`phenoData` and `featureData` slots show the content of each dataset. The
`loadExposome` function can also be used to construct an *ExposomeSet* from
loaded *data.frames* as input (see `?loadExposome`).

  Data of a *ExposomeSet* object can be retrieved with different accessors.
There are four basic functions `sampleNames`, `exposureNames`, `familyNames`
and `phenotypeNames` to extract the IDs of individuals, exposures, exposure
families and phenotypes.

```
> head(sampleNames(expo))
[1] "id001" "id002" "id003" "id004" "id005" "id006"


> head(exposureNames(expo))
[1] "AbsPM25" "As"       "BDE100" "BDE138" "BDE153" "BDE154"


> familyNames(expo)
 [1] "Air Pollutants"    "Metals"           "PBDEs"
 [4] "Organochlorines"   "Bisphenol A"      "Water Pollutants"
 [7] "Built Environment" "Cotinine"         "Home Environment"
[10] "Phthalates"        "Noise"            "PFOAs"
[13] "Temperature"


> phenotypeNames(expo)
[1] "whistling_chest" "flu"            "rhinitis"       "wheezing"
[5] "birthdate"       "sex"            "age"            "cbmi"
[9] "blood_pre"
```

Datasets are retrieved similarly to *ExpressionSet* objects. The `fData` function returns the description of the exposures

```
> head(fData(expo))
                      Family                                  Name .fct
AbsPM25 Air Pollutants Measurement of the blackness of PM2.5 filters
As                     Metals                                 Asenic
BDE100                  PBDEs        Polybrominated diphenyl ether -100
BDE138                  PBDEs        Polybrominated diphenyl ether -138
BDE153                  PBDEs        Polybrominated diphenyl ether -153
BDE154                  PBDEs        Polybrominated diphenyl ether -154
           .trn .std .imp   .type
AbsPM25                    numeric
As                         numeric
BDE100                     numeric
BDE138                     numeric
BDE153                     numeric
BDE154                     numeric
```

`pData` returns the phenotype data:

```
> head(pData(expo))
      whistling_chest flu rhinitis wheezing  birthdate    sex age cbmi
id001           never  no       no       no 2004-12-29   male 4.2 16.3
id002           never  no       no       no 2005-01-05   male 4.2 16.4
id003         7-12 epi  no       no      yes 2005-01-05   male 4.2 19.0
id004           never  no       no       no 2005-01-01 female 4.3 15.5
id005           never  no       no       no 2005-02-01 female 4.2 14.9
id006          1-2 epi  no      yes       no 2005-01-01   male 4.1 14.8
      blood_pre
id001       120
id002       121
id003       120
id004       117
id005       121
id006       113
```

and `expos` retrieves the matrix of exposures as a *data.frame*

```
> expos(expo)[1:10, c("Cotinine", "PM10V", "PM25", "X5cxMEPP")]
         Cotinine       PM10V     PM25 X5cxMEPP
id001  0.03125173  0.10373078 1.176255       NA
id002  1.59401990 -0.47768393 1.155122       NA
id003  1.46251090          NA 1.215834 1.859045
id004  0.89059991          NA 1.171610       NA
id005          NA          NA 1.145765       NA
id006  0.34818304          NA 1.145382       NA
id007  1.53591130          NA 1.174642       NA
id008  2.26864700          NA 1.165078 1.291871
id009  1.24842660          NA 1.171406 1.650948
id010 -0.36758339  0.01593277 1.179240 2.112357
```

3.8 *MultiAssayExperiment*

MultiAssayExperiment is a class designed to store the data of studies that collect multiple *omic* data on the same individuals [117]. A *MultiAssayExperiment* object, an extension of previous individual *omic* structures, has four data slots: ExperimentList, colData, sampleMap and metadata. `ExperimentList` contains the multiple *omic* data, each of which is any class supporting two dimensional structures, such as *matrix*, *SummarizedExperiment* or *ExpressionSet*. The slot `colData` stores the subjects' phenotype metadata that are common across the different *omic* datasets, such as the age or sex of the individuals. `sampleMap` links the names in `colData` to the column names in `ExperimentList` and `metadata` contains the study lab and experimental specifications.

As an example, we create a *MultiAssayExperiment* object from methylation and expression data available at `brgedata`. The function `MultiAssayExperiment` creates a new object, from the list of *omic* data, the subjects' metadata and the `sampleMap` object. If subjects have consistent names across all the different *omic* data, `sampleMap` is not required.

```
> library(MultiAssayExperiment)
> objlist <- list(expression = brge_gexp, methylation = brge_methy)
> mae <- MultiAssayExperiment(objlist)
> mae
A MultiAssayExperiment object of 2 listed
 experiments with user-defined names and respective classes.
 Containing an ExperimentList class object of length 2:
 [1] expression: ExpressionSet with 67528 rows and 100 columns
 [2] methylation: GenomicRatioSet with 392277 rows and 20 columns
Features:
 experiments() - obtain the ExperimentList instance
 colData() - the primary/phenotype DataFrame
 sampleMap() - the sample availability DataFrame
 `$`, `[`, `[[` - extract colData columns, subset, or experiment
 *Format() - convert into a long or wide DataFrame
 assays() - convert ExperimentList to a SimpleList of matrices
```

This particular example lacks annotation data, which could be added to the `colData` slot. For each slot, there is a function to access the data with the corresponding name

```
> experiments(mae)
ExperimentList class object of length 2:
 [1] expression: ExpressionSet with 67528 rows and 100 columns
 [2] methylation: GenomicRatioSet with 392277 rows and 20 columns
> colData(mae)
DataFrame with 100 rows and 0 columns
> sampleMap(mae)
DataFrame with 120 rows and 3 columns
         assay      primary      colname
```

```
          <factor> <character> <character>
1     expression       x0001       x0001
2     expression       x0002       x0002
3     expression       x0003       x0003
4     expression       x0004       x0004
5     expression       x0005       x0005
...          ...         ...         ...
116  methylation       x0018       x0018
117  methylation       x0057       x0057
118  methylation       x0061       x0061
119  methylation       x0077       x0077
120  methylation       x0079       x0079
> metadata(mae)
NULL
```

To obtain the *omic* data matrices, we use the function `assays`, that retrieves a list of matrices of each *omic* experiment:

```
> matrices <- assays(mae)
> names(matrices)
[1] "expression"  "methylation"
> matrices[[1]][1:4, 1:5]
                   x0001     x0002     x0003     x0004     x0005
TC01000001.hg.1 6.146656  5.852305  6.058288  5.922759  7.795354
TC01000002.hg.1 4.307691  4.017098  4.111963  4.118404  4.079342
TC01000003.hg.1 2.321697  2.570817  2.459935  2.421771  2.185509
TC01000004.hg.1 4.797603  4.314909  3.969750  5.178148  5.546576
> matrices[[2]][1:4, 1:5]
                x0017      x0043      x0036      x0041      x0032
cg13869341 0.90331003 0.90082827 0.8543610 0.84231205 0.92736037
cg24669183 0.86349463 0.84009629 0.8741255 0.85122181 0.85296304
cg15560884 0.65986032 0.69634338 0.6931083 0.69752294 0.70299219
cg01014490 0.01448404 0.01509478 0.0163614 0.01322362 0.01355301
```

MultiAssayExperiment objects can be subsetted by features, individuals or assays. Any subset is applied simultaneously to all assays across the entire data structure. We can select, for each dataset, a subset of individuals or of features that map to a specific genomic range. For instance, in our example we can select all the data for the two subjects with IDS x0001 and x0002

```
> mae[ , c("x0001", "x0002")]
A MultiAssayExperiment object of 1 listed
 experiment with a user-defined name and respective class.
 Containing an ExperimentList class object of length 1:
 [1] expression: ExpressionSet with 67528 rows and 2 columns
Features:
 experiments() - obtain the ExperimentList instance
 colData() - the primary/phenotype DataFrame
 sampleMap() - the sample availability DataFrame
 `$`, `[`, `[[` - extract colData columns, subset, or experiment
 *Format() - convert into a long or wide DataFrame
 assays() - convert ExperimentList to a SimpleList of matrices
```

3.9 *MultiDataSet*

MultiDataSet is another data structure that handles studies with different *omic* datasets [54]. A *MultiDataSet* object contains four data slots: the *omic* data (assayData), the individual's phenotypes (colData), the feature annotation (featureData) and a return function. Each slot is a *list* that corresponds to the different *omic* experiments.

The slots of *MultiDataSet* objects are first created empty with the function `createMultiDataSet` then the datasets are added. *MultiDataSet* includes functions to add common data classes, such as `ExpressionSet`, `SummarizedExperiment` or matrices. In the following example, we create a *MultiDataSet* object with the `brge_methy` and `brge_gexp` datasets that are available in *brgedata*

```
> data(brge_methy, package="brgedata")
> data(brge_gexp, pckage="brgedata")
>
> library(MultiDataSet)
> mds <- createMultiDataSet()
> mds <- add_methy(mds, brge_methy)
> mds <- add_genexp(mds, brge_gexp)
> mds
Object of class 'MultiDataSet'
 . assayData: 2 elements
    . methylation: 392277 features, 20 samples
    . expression: 67528 features, 100 samples
 . featureData:
    . methylation: 392277 rows, 19 cols (seqnames, ..., Regulatory_Feature_G)
    . expression: 67528 rows, 11 cols (transcript_cluster_id, ..., gene_assig)
 . rowRanges:
    . methylation: YES
    . expression: YES
 . phenoData:
    . methylation: 20 samples, 10 cols (age, ..., Neu)
    . expression: 100 samples, 3 cols (age, ..., sex)
```

`add_methy` and `add_genexp` add methylation and gene expression data to an initial *MultiDataSet* object, respectively. These functions ensure that added data fulfill transcriptomic and methylomic requirements, adding a tag to each specific *omic* data. As a result, methods can recognize the expected structure of the data to apply the correct functions.

`MultiDataSet` can be subsetted by features, individuals or assays. Any subset is applied simultaneously to all assays and to all sample and feature annotations. For instance, we can select all features mapped to chromosome 1, using a *GRanges* object:

```
> range <- GRanges("chr1:1-999999999")
> mds[, , range]
Object of class 'MultiDataSet'
```

```
. assayData: 2 elements
  . methylation: 38395 features, 20 samples
  . expression: 6342 features, 100 samples
. featureData:
  . methylation: 38395 rows, 19 cols (seqnames, ..., Regulatory_Feature_G)
  . expression: 6342 rows, 11 cols (transcript_cluster_id, ..., gene_assig)
. rowRanges:
  . methylation: YES
  . expression: YES
. phenoData:
  . methylation: 20 samples, 10 cols (age, ..., Neu)
  . expression: 100 samples, 3 cols (age, ..., sex)
```

We can also select data from a subset of samples:

```
> mds[c("x0001", "x0002"), ]
Object of class 'MultiDataSet'
. assayData: 2 elements
  . methylation: 392277 features, 0 samples
  . expression: 67528 features, 2 samples
. featureData:
  . methylation: 392277 rows, 19 cols (seqnames, ..., Regulatory_Feature_G)
  . expression: 67528 rows, 11 cols (transcript_cluster_id, ..., gene_assig)
. rowRanges:
  . methylation: YES
  . expression: YES
. phenoData:
  . methylation: 0 samples, 10 cols (age, ..., Neu)
  . expression: 2 samples, 3 cols (age, ..., sex)
```

MultiDataSet also has a predefined function (commonSamples) to return a
MultiDataSet object with complete cases, i.e. an object with common samples
across all *omic* datasets can be retrieved

```
> commonSamples(mds)
Object of class 'MultiDataSet'
. assayData: 2 elements
  . methylation: 392277 features, 20 samples
  . expression: 67528 features, 20 samples
. featureData:
  . methylation: 392277 rows, 19 cols (seqnames, ..., Regulatory_Feature_G)
  . expression: 67528 rows, 11 cols (transcript_cluster_id, ..., gene_assig)
. rowRanges:
  . methylation: YES
  . expression: YES
. phenoData:
  . methylation: 20 samples, 10 cols (age, ..., Neu)
  . expression: 20 samples, 3 cols (age, ..., sex)
```

We illustrate now how to create a MultiDataSet object from a real data
example. It corresponds to a study on breast cancer available at TCGA. Data
was downloaded from https://tcga-data.nci.nih.gov/docs/publica
tions/brca_2012/ and saved in an R object called breastMulti_list,
which is available through the *brgedata* package. The object is a list containing miRNA, miRNAprecursor, RNAseq, Methylation, proteins from a RPPA

array, and GISTIC SNP calls (CNA and LOH). Clinical data are also available. The following code shows how to create a `MultiDataSet` object for these data, from `GenomicRatioSet` and `SummarizedExperiment` objects and their required feature annotation.

We first list the different omic tables as loaded into R by the *brgedata* package

```
> data(breastMulti_list, package="brgedata")
```

Each component of the list contains different *omic* data with the following number of features (rows) and samples (columns):

```
> lapply(breastMulti_list, dim)
$miRNA
[1] 929  79

$miRNAprecursor
[1] 772  79

$RNAseq
[1] 10020    79

$Methyl
[1] 574  79

$RPPA
[1] 171  79

$LOH
[1] 12612    79

$CNA
[1] 11831    79

$clin
[1] 79 29
```

We then create a `GenomicRatioSet` object for the data frame containing methylation data (see Section 3.4). This can be done with the function `makeGenomicRatioSetFromMatrix` from the *minfi* package

```
> library(minfi)
> methyBreast <- breastMulti_list$Methy
> methyBreast <- makeGenomicRatioSetFromMatrix(data.matrix(methyBreast))
> class(methyBreast)
[1] "GenomicRatioSet"
attr(,"package")
[1] "minfi"
```

We now create a `SummarizedExperiment` object for gene expression data. While compatible with `ExpressionSet` objects, `SummarizedExperiment` structure is currently preferred. `SummarizedExperiment` objects first require annotation data, which we obtain from the `biomaRt` package and store in `annot`. The names of the gene features are obtained from the rownames of the expression data `breastMulti_list$RNAseq`:

```
> library(biomaRt)
> ensembl <- useMart("ensembl")
> ensembl <- useDataset("hsapiens_gene_ensembl", mart=ensembl)
> symbol.genes <- rownames(breastMulti_list$RNAseq)
> annot <- getBM(attributes=c('hgnc_symbol',
+                             'chromosome_name',
+                             'start_position',
+                             'end_position'),
+            filters=c('hgnc_symbol'),
+            values=symbol.genes,
+            mart=ensembl)
```

Duplicated genes are removed and annotation is converted into a `GenomicRanges` object:

```
> names(annot) <- c("symbol", "chromosome", "start", "end")
> annot$chromosome <- paste0("chr", annot$chromosome)
> annot <- annot[!duplicated(annot[,1]),]
> rownames(annot) <- annot$symbol
> annot <- GenomicRanges::makeGRangesFromDataFrame(annot)
```

Finally, we create the `SummarizedExperiment` object with the annotated genes as follows

```
> genes <- intersect(names(annot), symbol.genes)
> exprBreast <- breastMulti_list$RNAseq[genes, ]
> exprBreast <- SummarizedExperiment(assays=list(genexpr=exprBreast),
+                                    rowRanges=annot[genes,])
> exprBreast
class: RangedSummarizedExperiment
dim: 8297 79
metadata(0):
assays(1): genexpr
rownames(8297): ABCA6 ABCC12 ... ZNF569 ZNF610
rowData names(0):
colnames(79): TCGA-C8-A12V TCGA-A2-A0ST ... TCGA-B6-A0X0
  TCGA-A2-A04Y
colData names(0):
```

Other *omic* data can be formatted as `ExpressionSet` objects

```
> miRNABreast <- ExpressionSet(data.matrix(breastMulti_list$miRNA))
> miRNApreBreast <- ExpressionSet(data.matrix(breastMulti_list$miRNAprecursor))
> proteBreast <- ExpressionSet(data.matrix(breastMulti_list$RPPA))
> lohBreast <- ExpressionSet(breastMulti_list$LOH)
> cnaBreast <- ExpressionSet(breastMulti_list$CNA)
```

We now create the phenotypic data structure (e.g. clinical data) for the different *omic* objects. Remember that `colData` is used for `SummarizedExperiment` objects and pData for `ExpressionSet`.

```
> breastMulti_list$clin <- droplevels(breastMulti_list$clin)
>
> colData(methyBreast) <- colData(exprBreast) <-
+   DataFrame(breastMulti_list$clin)
> pData(miRNABreast) <- pData(miRNApreBreast) <- pData(proteBreast) <-
+   pData(lohBreast) <- pData(cnaBreast) <- breastMulti_list$clin
```

The **MultiDataSet** object for the multi-*omic* experiment is then created by attaching all previous data structures as follow

```
> library(MultiDataSet)
> breastMulti <- createMultiDataSet()
> breastMulti <- add_methy(breastMulti, methyBreast)
> breastMulti <- add_rse(breastMulti, exprBreast, "expression")
> breastMulti <- add_eset(breastMulti, miRNABreast, "miRNA",
+                         GRanges = NA)
> breastMulti <- add_eset(breastMulti, miRNApreBreast, "miRNAprecursor",
+                         GRanges = NA)
> breastMulti <- add_eset(breastMulti, proteBreast, "RPPA",
+                         GRanges = NA)
> breastMulti <- add_eset(breastMulti, lohBreast, "LOH",
+                         GRanges = NA)
> breastMulti <- add_eset(breastMulti, cnaBreast, "CNA",
+                         GRanges = NA)
> breastMulti
Object of class 'MultiDataSet'
 . assayData: 7 elements
    . methylation: 574 features, 79 samples
    . expression: 8297 features, 79 samples
    . miRNA: 929 features, 79 samples
    . miRNAprecursor: 772 features, 79 samples
    . RPPA: 171 features, 79 samples
    . LOH: 12612 features, 79 samples
    . CNA: 11831 features, 79 samples
 . featureData:
    . methylation: 574 rows, 5 cols (seqnames, ..., width)
    . expression: 8297 rows, 5 cols (seqnames, ..., width)
    . miRNA: 929 rows, 0 cols
    . miRNAprecursor: 772 rows, 0 cols
    . RPPA: 171 rows, 0 cols
    . LOH: 12612 rows, 0 cols
    . CNA: 11831 rows, 0 cols
 . rowRanges:
    . methylation: YES
    . expression: YES
    . miRNA: NO
    . miRNAprecursor: NO
    . RPPA: NO
    . LOH: NO
    . CNA: NO
 . phenoData:
    . methylation: 79 samples, 30 cols (Gender, ..., Integrated.Clusters..)
    . expression: 79 samples, 30 cols (Gender, ..., Integrated.Clusters..)
    . miRNA: 79 samples, 30 cols (Gender, ..., Integrated.Clusters..)
    . miRNAprecursor: 79 samples, 30 cols (Gender, ..., Integrated.Clusters..)
    . RPPA: 79 samples, 30 cols (Gender, ..., Integrated.Clusters..)
```

```
. LOH: 79 samples, 30 cols (Gender, ..., Integrated.Clusters..)
. CNA: 79 samples, 30 cols (Gender, ..., Integrated.Clusters..)
```

4

Genetic association studies

CONTENTS

4.1 Chapter overview

Genetic association studies aim to identify genetic variants that explain subject differences in qualitative or quantitative traits. Genetic variants may include SNPs, CNVs, mosaicisms or polymorphic inversions, which can be obtained from SNP array data. In this chapter, we thoroughly discuss SNP association studies. We explain the different genetic models to be used in single SNP associations and how they are applied in a massive univariate testing when several SNPs are analyzed. Relating the simultaneous analysis of multiple SNPs, we discuss haplotype and genetic scores associations. We then move to the analysis of genome-wide association studies, where high-dimensional genomic data is treated. Issues concerning quality control of SNPs and individuals are discussed, as well as association tests and possible sources of confounding such as population stratification. We finish SNP analyses with tools that help to interpret results.

4.2   Genetic association studies

In this section, we illustrate how to perform SNP association analyses. We discuss single SNP association analyses, multiple SNPs analyses (haplotype analysis and genetic scores), environmental interactions and genome-wide association analyses (GWAS). We cover quality control, population stratification and multiple comparisons issues. We also offer guidelines for visualization and annotation of results.

4.2.1   Analysis packages

There are several R/Bioconductor packages to analyze genetic association studies. *gap* is a package for the analysis of population and family data; it contains functions for sample size calculations, the probability of familial disease aggregation, kinship and tests for linkage and association analyses. For family data, *tdthap* offers an implementation of the Transmission/Disequilibrium Test (TDT) for extended marker haplotypes.

One of the most widely used packages to perform genetic association studies is *SNPassoc* that contains a complete analysis toolkit, which includes data manipulation, exploratory data analysis, and visualization [42]. Within *SNPassoc*, one can assess the genetic association of quantitative and categorical (case-control) traits for targeted SNPs or SNPs in genomic data. Different genetic models are implemented in the package. Multiple SNPs can be analyzed using either haplotype or genetic scores and analyses for gene-environment and gene-gene interactions of candidate SNPs are also available. We will demonstrate the use of *SNPassoc*.

Although *SNPassoc* can deal with thousands of SNPs, there are other R and Bioconductor packages that are more efficient in the analysis of high-dimensional SNP data. *GenABEL* and *snpStats* are designed for the efficient storage and handling of GWAS data with fast tools for quality control, association with binary and quantitative traits, as well as tools for visualizing results. One of the main advantages of *snpStats* is its efficiency in loading and handling data in PLINK format, helping to reduce computing time. *snpStats* also includes an efficient implementation of principal component analysis and Q-Q plots. One of its main drawbacks is that it only considers additive genetic models and does not offer Manhattan plots. Nonetheless, its results can be processed with other R/Bioconductor tools that are particularly designed for post-GWAS analyses.

4.2.2 Association tests

A single nucleotide polymorphism (SNP) is a categorical variable that encodes one of three possible genotypes (homozygous normal: A/A, heterozygous: A/a, and homozygous variant: a/a) per individual at a given genomic locus. We are interested in testing whether a SNP is significantly associated with a quantitative trait across subjects. The statistical test to assess the association between two categorical variables is the χ^2 test where the strength of the association is given by the odds ratio (OR). Figure 4.1 illustrates the association between a given SNP and a case-control status of individuals. The first table represent the codominant model, where each SNP genotype confers a significantly different risk. In the codominant model, also known as multiplicative or model-free, the ORs are computed for A/a ($OR_{A/a}$) and a/a ($OR_{a/a}$) with respect to A/A individuals. The additive genetic model, each alternative allele a confers an independent additive risk and, therefore, the ($OR_{a/a}$) is twice/half ($OR_{A/a}$) in log-scale. The main difference between codominant and additive models is that the first has 2 degrees of freedom (d.f.) while the second 1 d.f. and, consequently, the additive model has more power to detect differences than codominant model when linear trend can be assumed.

The association analysis can also be performed assuming a dominant genetic model, where the presence of at least one variant allele fully determines the phenotypic differences between individuals (Figure 4.1). The analysis is performed by aggregating the A/a and a/a individuals into a single category in the contingency table. On the other hand, the association analyses using a recessive model is performed by aggregating the A/A and A/a individuals into a unique category, in which two variant alleles are needed to confer phenotypic differences. In both cases a χ^2 test is used to assess association on the resulting 2×2 tables. If SNPs have a low frequency for the variant allele then a Fisher's test may be required.

When multiple SNPs are analyzed independently, each may follow a different genetic model. It is a usual practice to consider testing all associations using the additive model, which is the most powerful in practice [6, 80]. Another option is to consider dominant, recessive and additive models for each SNP and declare a significant association when the minimum of the three P-values pass a predetermined significance level. The level must be lower than the one that is applied in a single significant test (i.e. nominal level 0.05), since the probability to reject the null hypothesis by chance, when three tests are applied, increases. To redefine the significance level, that accounts for these multiple testings, one could apply Bonferroni correction, assuming that the three tests are independent, and divide the nominal significance level by 3 (i.e.

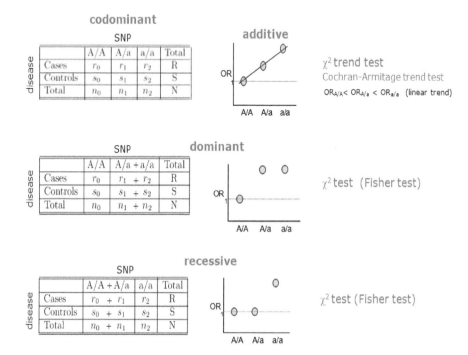

FIGURE 4.1

Genetic models and statistical tests used in assessing single association between a discrete trait (disease) and a given SNP. The number of cases for
each genotype are denoted by r_0, r_1 and r_2 while for controls are s_0, s_1, s_2. A
given SNP is associated with the disease when the proportion of each genotype
differs between the total number of cases (R) and controls (N). The codomi-
nat model tests the global homogeneity of those proportions, while the other
genetic models test diffent assumtions (additive: linearity, dominant: risk of a
allele, recessive: risk of a/a genotype).

0.05/3). However, the three models are not independent and the significance level is better determined from a max-statistic test, which is derived from the three χ^2 statistics (1 d.f.) for each model [43]. As a rule of thumb, one can consider that the number of independent tests is 2.2, and consider the significance level at 0.0227 for a nominal level of 0.05 (0.0227 = 0.05/2.2) [43, 67]. The *P*-value of the max-statistic is implemented in *SNPassoc*.

In the next sections, we demonstrate the use of *SNPassoc* to perform genetic associations using different genetic models and a max-statistic test, haplotype and genetic score analyses, and gene-gene and gene-environment interactions. GWASs including quality control, population stratification and multiple comparisons will be demonstrated with Bioconductor's packages *SNPRelate* and *snpStats*. Post-GWAS visualization is illustrated using the *Gviz* package.

4.2.3 Single SNP analysis

We describe the analysis of genetic association studies that are based on a moderate number of targeted SNPs. The SNPs are typically selected from genomic regions that have a hypothetical role in the development of a trait of interest. For instance, SNPs may be selected from candidate genes, or are a set of neighboring SNPs that finely map a candidate genomic region linked to a disease. In addition, one may be interested in studying a given set of SNPs to validate previously reported findings and determine the best possible genetic model of inheritance (codominant, dominant, recessive, additive).

SNP data for this type of experiments are typically available in text format or Excel spreadsheets which are easily uploaded in R. A dataset containing epidemiological information and 51 SNPs from a case-control study on asthma can be obtained from the *brgedata*, introduced in Chapter 3.

```
> data(asthma, package = "brgedata")
> str(asthma, list.len=9)
'data.frame': 1578 obs. of  57 variables:
 $ country    : Factor w/ 10 levels "Australia","Belgium",..
                : 5 5 5 5 5 5 5 5 5 ...
 $ gender     : Factor w/ 2 levels "Females","Males": 2 2 2 1 1 1 1 2 1 1 ...
 $ age        : num   42.8 50.2 46.7 47.9 48.4 ...
 $ bmi        : num   20.1 24.7 27.7 33.3 25.2 ...
 $ smoke      : int   1 0 0 0 0 1 0 0 0 0 ...
 $ casecontrol: int   0 0 0 0 1 0 0 0 0 0 ...
 $ rs4490198  : Factor w/ 3 levels "AA","AG","GG": 3 3 3 2 2 2 3 2 2 2 ...
 $ rs4849332  : Factor w/ 3 levels "GG","GT","TT": 3 2 3 2 1 2 3 3 2 1 ...
 $ rs1367179  : Factor w/ 3 levels "CC","GC","GG": 2 2 2 3 3 3 2 3 3 3 ...
  [list output truncated]
> asthma[1:5, 1:10]
   country  gender       age      bmi smoke casecontrol rs4490198 rs4849332
1  Germany   Males 42.80630 20.14797     1           0        GG        TT
2  Germany   Males 50.22861 24.69136     0           0        GG        GT
3  Germany   Males 46.68857 27.73230     0           0        GG        TT
4  Germany Females 47.86311 33.33187     0           0        AG        GT
```

```
5 Germany Females 48.44079 25.23634     0              1          AG        GG
  rs1367179 rs11123242
1         GC         CT
2         GC         CT
3         GC         CT
4         GG         CC
5         GG         CC
```

We observe that we have case-control status (0: control, 1: asthma) and another 4 variables encoding the country of origin, gender, age, body mass index (bmi) and smoking status (0: no smoker, 1: ex-smoker, 2: current smoker). There are 51 SNPs whose genotypes are given by the alleles names. We analyze this data with *SNPassoc* package. The package can be downloaded from CRAN, with the command `install.packages("SNPassoc")`. The development version that includes new features and functions is available through GitHub and can be installed using the *devtools* package

```
> require(devtools)
> devtools::install_github("isglobal-brge/SNPassoc")
```

To start the analysis, we must indicate which columns of the dataset `asthma` contain the SNP data, using the `setupSNP` function. In our example, SNPs start from column 7 onwards, which we specify in argument `colSNPs`

```
> library(SNPassoc)
> asthma.s <- setupSNP(data=asthma, colSNPs=7:ncol(asthma), sep="")
```

The argument `sep` indicates the character separating the alleles. The default value is "/. In our case, there is no separating character and we set `sep=""`. The argument `name.genotypes` can be used when genotypes are available in other formats, such as 0, 1, 2 or "norm, "het, "mut. The purpose of the `setupSNP` function is to assign the class *snp* to the SNPs variables, to which *SNPassoc* methods will be applied. The function labels the most common genotype across subjects as the reference one. When numerous SNPs are available, the function can be parallelized through the argument `mc.cores` that indicates the number of processors to be used. We can verify that the SNP variables are given the new class *snp*

```
> head(asthma.s$rs1422993)
[1] G/G G/T G/G G/T G/T G/G
Genotypes: G/G G/T T/T
Alleles:  G T
> class(asthma.s$rs1422993)
[1] "snp"    "factor"
```

and summarize their content with `summary`

```
> summary(asthma.s$rs1422993)
Genotypes:
    frequency percentage
```

```
G/G      903  57.224335
G/T      570  36.121673
T/T      105   6.653992

Alleles:
  frequency percentage
G      2376   75.28517
T       780   24.71483

HWE (p value): 0.250093
```

which shows the genotype and allele frequencies for a given SNP, testing for
Hardy–Weinberg equilibrium (HWE), explained below. The **summary** function
can also be applied to the whole dataset

```
> ss <- summary(asthma.s, print=FALSE)
> head(ss)
            alleles major.allele.freq     HWE missing
rs4490198      A/G               59.2 0.174133    0.6
rs4849332      G/T               61.8 0.522060    0.1
rs1367179      G/C               81.4 0.738153    1.0
rs11123242     C/T               81.7 0.932898    0.6
rs13014858     G/A               58.3 0.351116    0.1
rs1430094      G/A               66.9 0.305509    0.4
```

showing the SNP labels with minor/major allele format, the major allele frequency the HWE test and the percentage of missing genotypes. Missing values
can be further explored plotting

```
> plotMissing(asthma.s, print.labels.SNPs = FALSE)
```

Figure 4.2 shows the missing genotype pattern across SNPs. This plot can
be used to inspect if missing values appear randomly across individuals and
SNPs. In our case, we can see that the missing pattern may be considered
random, except for three clusters in consecutive SNPs (large black squares).
These individuals should be further checked for possible problems with genotyping.

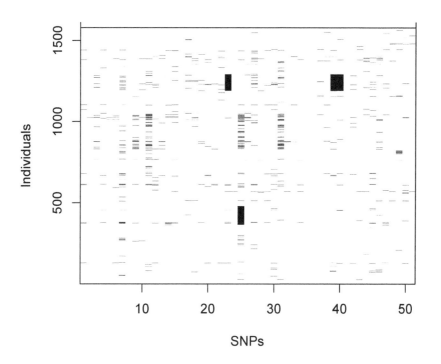

FIGURE 4.2
Missing genotypes (black) of asthma data example.

4.2.4 Hardy–Weinberg equilibrium

Genotyping of SNPs needs to pass quality control measures. Aside from technical details that need to be considered for filtering SNPs with low quality, genotype calling error can be detected by a Hardy–Weinberg equilibrium (HWE) test. The test compares the observed genotype frequencies with those expected under random mating, which follows when the SNPs are in the absence of selection, mutation, genetic drift, or other forces. Therefore, HWE must be checked only in controls. There are several tests described in the literature to verify HWE. For instance, in *SNPassoc*, HWE is tested for all the bi-allelic SNP markers using a fast exact test [169]. HWE can be specifically tested with the `tableHWE` function.

```
> hwe <- tableHWE(asthma.s)
> head(hwe)
            HWE (p value)
rs4490198       0.1741325
rs4849332       0.5220596
rs1367179       0.7381531
rs11123242      0.9328981
rs13014858      0.3511162
rs1430094       0.3055089
```

We observe that the first SNPs in the dataset are under HWE since their *P*-values rejecting the HWE hypothesis (null hypothesis) are larger than 0.05. However, when tested in control samples only, by stratifying by cases and controls

```
> hwe2 <- tableHWE(asthma.s, casecontrol)
>
> #SNPs is HWE in the whole sample but not controls
> snpNHWE <- hwe2[,1]>0.05 & hwe2[,2]<0.05
> rownames(hwe2)[snpNHWE]
[1] "rs1345267"
> hwe2[snpNHWE,]
all groups          0          1
0.11218285 0.04956349 0.81604706
```

we see that rs1345267 is not in HWE within controls because its *P*-value is < 0.05. Notice that one is interested in keeping those SNPs that do not reject the null hypothesis. As several SNPs are tested, multiple comparisons must be considered. In this particular setting, a threshold of 0.001 is normally considered. As a quality control measure, it is not necessary to be as conservative as in those situations where false discovery rates need to be controlled.

SNPs that do not pass the HWE test must be removed form further analyses. We can recall **setupSNP** and indicate the columns of the SNPs to be kept

```
> snps.ok <- rownames(hwe2)[hwe2[,2]>=0.001]
> pos <- which(colnames(asthma)%in%snps.ok, useNames = FALSE)
> asthma.s <- setupSNP(asthma, pos, sep="")
```

in the variable pos, we redefine the SNP variables to be considered as class *snp*.

4.2.5 SNP association analysis

We are interested in finding those SNPs associated with asthma status that is encoded in the variable casecontrol. We first illustrate the association between case-control status and the SNP rs1422993. The association analysis for all genetic models is performed by the function association that regresses casecontrol on the variable rs1422993 of class *snp* in the dataset asthma.s

```
> association(casecontrol ~ rs1422993, data = asthma.s)

SNP: rs1422993   adjusted by:
                   0    %    1    %   OR lower upper  p-value  AIC
Codominant
G/G              730 59.0 173 50.9 1.00                0.017768 1642
G/T              425 34.3 145 42.6 1.44  1.12  1.85
T/T               83  6.7  22  6.5 1.12  0.68  1.84
Dominant
G/G              730 59.0 173 50.9 1.00                0.007826 1642
G/T-T/T          508 41.0 167 49.1 1.39  1.09  1.77
Recessive
G/G-G/T         1155 93.3 318 93.5 1.00                0.877863 1649
T/T               83  6.7  22  6.5 0.96  0.59  1.57
Overdominant
G/G-T/T          813 65.7 195 57.4 1.00                0.005026 1641
G/T              425 34.3 145 42.6 1.42  1.11  1.82
log-Additive
0,1,2           1238 78.5 340 21.5 1.22  1.01  1.47 0.040151 1644
```

association follows the usual syntax of lm and glm with the difference that the variables in the model that are of class *snp* are tested using different genetic models. In our example, we observe that all genetic models but the recessive one are statistically significant. association also fits the overdominant model, which compares the two homozygous genotypes versus the heterozygous one. This genetic model of inheritance is biologically rare although it has been linked to sickle cell anemia in humans. The result table describes the number of individuals in each genotype across cases and controls. The ORs and CI-95% are also computed. The last column describes the AIC (Akaike information criteria) that can be used to decide which is the best model of inheritance; the lower the better the model is. In the example, one may conclude that rs1422993 is associated with asthma and that, for instance, the risk of being asthmatic is 39% higher in people having at least one alternative allele (T) with respect to individuals having none (dominant model). This risk is statistically significant since the CI-95% does not contain 1, the P-value is $0.0078 < 0.022$. or the P-value of the max-statistics is 0.01

```
> maxstat(asthma.s$casecontrol, asthma.s$rs1422993)
     dominant recessive log-additive MAX-statistic Pr(>z)
[1,]    7.073      0.024        4.291         7.073 0.0143 *
---
Signif. codes:  0 '***' 0.001 '**' 0.01 '*' 0.05 '.' 0.1 ' ' 1
```

If an expected model of inheritance is hypothesized, the association analysis for the model can be specified in the argument `model`, which by default test all models,

```
> association(casecontrol ~ rs1422993, asthma.s, model="dominant")
```

```
SNP: rs1422993  adjusted by:
             0  %  1   %  OR lower upper  p-value  AIC
Dominant
G/G       730 59 173 50.9 1.00          0.007826 1642
G/T-T/T   508 41 167 49.1 1.39  1.09  1.77
```

Association tests are typically adjusted by covariates, which are incorporated in the model in the usual form

```
> association(casecontrol ~ rs1422993 + country + smoke, asthma.s)
```

```
SNP: rs1422993  adjusted by: country smoke
                0   %  1    %  OR lower upper p-value  AIC
Codominant
G/G          728 59.1 173 51.0 1.00          0.06940 1408
G/T          423 34.3 144 42.5 1.38  1.05  1.82
T/T           81  6.6  22  6.5 1.06  0.61  1.85
Dominant
G/G          728 59.1 173 51.0 1.00          0.03429 1407
G/T-T/T      504 40.9 166 49.0 1.33  1.02  1.73
Recessive
G/G-G/T     1151 93.4 317 93.5 1.00          0.79338 1411
T/T           81  6.6  22  6.5 0.93  0.54  1.60
Overdominant
G/G-T/T      809 65.7 195 57.5 1.00          0.02147 1406
G/T          423 34.3 144 42.5 1.37  1.05  1.80
log-Additive
0,1,2       1232 78.4 339 21.6 1.19  0.96  1.46 0.11191 1409
```

ORs for stratified analysis on given categorical covariates are used to verify whether the risk is constant across groups

```
> association(casecontrol ~ rs1422993 + survival::strata(gender), asthma.s)
```

```
SNP: rs1422993  adjusted by: survival::strata(gender)
                0   %  1    %  OR lower upper  p-value  AIC
Codominant
G/G          730 59.0 173 50.9 1.00          0.022940 1634
G/T          425 34.3 145 42.6 1.42  1.11  1.83
T/T           83  6.7  22  6.5 1.09  0.66  1.80
Dominant
G/G          730 59.0 173 50.9 1.00          0.011144 1633
```

```
G/T-T/T          508 41.0 167 49.1 1.37  1.07  1.74
Recessive
G/G-G/T         1155 93.3 318 93.5 1.00              0.805330 1640
T/T               83  6.7  22  6.5 0.94  0.58  1.53
Overdominant
G/G-T/T          813 65.7 195 57.4 1.00              0.006378 1632
G/T              425 34.3 145 42.6 1.41  1.10  1.80
log-Additive
0,1,2           1238 78.5 340 21.5 1.21  1.00  1.46 0.055231 1636
```

We can see, for instance, that the dominant model is significant only in males. The subset argument allows fitting the model in a subgroup of individuals

```
> association(casecontrol ~ rs1422993, asthma.s,
+                  subset=country=="Spain")

SNP: rs1422993  adjusted by:
                   0   %  1   %   OR lower upper p-value   AIC
Codominant
G/G              179 54.6 22 44.9 1.00              0.3550 295.2
G/T              125 38.1 24 49.0 1.56  0.84  2.91
T/T               24  7.3  3  6.1 1.02  0.28  3.66
Dominant
G/G              179 54.6 22 44.9 1.00              0.2059 293.7
G/T-T/T          149 45.4 27 55.1 1.47  0.81  2.70
Recessive
G/G-G/T          304 92.7 46 93.9 1.00              0.7576 295.2
T/T               24  7.3  3  6.1 0.83  0.24  2.85
Overdominant
G/G-T/T          203 61.9 25 51.0 1.00              0.1502 293.2
G/T              125 38.1 24 49.0 1.56  0.85  2.85
log-Additive
0,1,2            328 87.0 49 13.0 1.23  0.77  1.96  0.3816 294.5
```

These analyses can be also be performed in quantitative traits, such as body mass index, as **association** automatically selects the error distribution of the regression analysis (either Gaussian or binomial).

```
> association(bmi ~ rs1422993, asthma.s)

SNP: rs1422993  adjusted by:
                  n   me    se      dif     lower  upper p-value  AIC
Codominant
G/G             896 25.53 0.1446  0.000000              0.9069 9069
G/T             565 25.50 0.1834 -0.027059 -0.4874 0.4332
T/T             105 25.71 0.4676  0.178076 -0.7057 1.0619
Dominant
G/G             896 25.53 0.1446  0.000000              0.9818 9067
G/T-T/T         670 25.54 0.1710  0.005089 -0.4324 0.4426
Recessive
G/G-G/T        1461 25.52 0.1135  0.000000              0.6694 9067
T/T             105 25.71 0.4676  0.188540 -0.6769 1.0540
Overdominant
```

```
G/G-T/T      1001 25.55 0.1383  0.000000                       0.8424 9067
G/T           565 25.50 0.1834 -0.045739 -0.4965 0.4050
log-Additive
0,1,2                            0.033951 -0.3153 0.3832  0.8489 9067
```

For BMI, `association` tests whether the difference between means is statistically significant, rather than computing an OR.

For multiple SNP data, our objective is to identify the variants that are significantly associated with the trait. The most basic strategy is, therefore, to fit an association test like the one described above for each of the SNPs in the dataset and determine which of those associations are significant. The massive univariate testing is the most widely used analysis method for *omic* data because of its simplicity. In *SNPassoc*, this type of analysis is done with the `WGassociation`

```
> ans <- WGassociation(casecontrol, data=asthma.s)
> head(ans)
            comments codominant dominant recessive overdominant
rs4490198       -       0.52765  0.29503   0.96400      0.29998
rs4849332       -       0.96912  0.92986   0.84806      0.82327
rs1367179       -       0.62775  0.59205   0.35786      0.86419
rs11123242      -       0.68622  0.67596   0.39801      0.92878
rs13014858      -       0.52578  0.26739   0.88011      0.34966
rs1430094       -       0.13375  0.10569   0.54432      0.04490
            log-additive
rs4490198       0.49506
rs4849332       0.97049
rs1367179       0.43994
rs11123242      0.52009
rs13014858      0.40897
rs1430094       0.36611
```

Here, only the outcome is required in the formula argument (first argument) since the function successively calls `association` on each of the variables of class *snp* within `data`. The function returns the *P*-values of association of each SNP under each genetic model. Covariates can also be introduced in the model

```
> ans.adj <- WGassociation(casecontrol ~ country + smoke, asthma.s)
> head(ans.adj)
```

SNPassoc is computationally limited on large *omic* data. The computing time can be reduced by parallelization, specifying in the argument `mc.cores` the number of computing cores to be used. Alternatively, the function `scanWGassociation`, a C compiled function, can be used to compute a predetermined genetic model across all SNPs, passed in the argument `model`, which by default is the additive model

```
> ans.fast <- scanWGassociation(casecontrol, asthma.s)
```

The *P*-values obtained from massive univariate analyses are visualized with the generic `plot` function

```
> plot(ans)
```

This produces a Manhattan plot of the $-\log_{10}(P)$-values for all the SNPs over all models. It shows the nominal level of significance and the Bonferroni level, which is the level corrected by the multiple testing across all SNPs. The overall hypothesis of massive univariate association tests is whether there is SNP *any* that is significantly associated with the phenotype. As multiple SNPs are tested, the probability of finding at least one significant finding increases if we do not lower the significance threshold. The Bonferroni correction lowers the threshold by the number of SNPs tested (0.0001=0.05/51). In the Manhattan plot of our analysis, we see that no SNP is significant at the Bonferroni level, and therefore there is no SNP that is significantly associated with asthma.

Maximum-statistics can be computed across all SNPs

```
> ans.max <- maxstat(asthma.s, casecontrol)
> head(ans.max)
[1] 1.096514182 0.002036721 0.466455726 1.096514182 0.502701763 0.007748582
```

Previous results on the maximum-statistic for rs1422993 can be recovered

```
> #Maximum-statistics for rs1422993
> ans.max[,"rs1422993"]
    dominant    recessive  log-additive MAX-statistic      Pr(>z)
  7.07284771   0.02361711    4.29114971    7.07284771   0.01854442
```

We note that even under the max-statistics none of the SNPs tested is significant under the Bonferroni correction (<0.0001) for multiple SNP testing

```
> #minimum P-value across SNPs
> min(ans.max["Pr(>z)",])
[1] 0.003288909
```

Information for specific association models for given SNPs can also be retrieved with `WGstats`

```
> infoTable <- WGstats(ans)
> infoTable$rs1422993
```

recovering our previous results given by `association`. The R output of specific association analyses can be exported into LaTeX by using `getNiceTable` function and *xtable* R package. The following code creates a table for the SNPs rs1422993 and rs184448

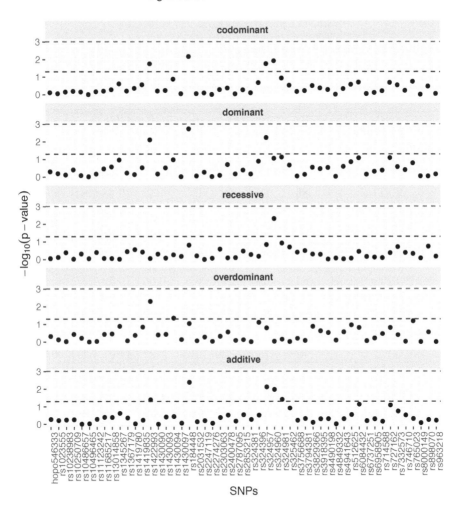

FIGURE 4.3

Manhattan-type plots for different genetic models to assess the association between case-control status and SNPs in the asthma example.

```
> library(xtable)
> out <- getNiceTable(ans[c("rs1422993", "rs184448")])
>
> nlines <- attr(out, "nlines")
> hlines <- c(-1, -1, 0, cumsum(nlines+1), nrow(out), nrow(out))
>
> print(xtable(out, caption='Genetic association using
+                  different genetic models from asthma
+                  data example of rs1422993 and rs184448
+                  SNPs obtained with SNPassoc.',
+              label = 'tab-2SNPs'),
+       tabular.enviroment="longtable", file="tableSNPs",
+       floating=FALSE,  include.rownames = FALSE,
+       hline.after= hlines, sanitize.text.function=identity)
```

SNP	0	%	1	%	OR	CI95%	p-value
rs1422993							
Codominant							
G/G	730	59.0	173	50.9	1.00		0.01777
G/T	425	34.3	145	42.6	1.44	(1.12-1.85)	
T/T	83	6.7	22	6.5	1.12	(0.68-1.84)	
Dominant							
G/G	730	59.0	173	50.9	1.00		0.007826
G/T-T/T	508	41.0	167	49.1	1.39	(1.09-1.77)	
Recessive							
G/G-G/T	1155	93.3	318	93.5	1.00		0.8779
T/T	83	6.7	22	6.5	0.96	(0.59-1.57)	
Overdominant							
G/G-T/T	813	65.7	195	57.4	1.00		0.005026
G/T	425	34.3	145	42.6	1.42	(1.11-1.82)	
log-Additive							
0,1,2	1238	78.5	340	21.5	1.22	(1.01-1.47)	0.04015
rs184448							
Codominant							
T/T	381	31.5	76	22.8	1.00		0.006777
T/G	624	51.5	189	56.8	1.52	(1.13-2.04)	
G/G	206	17.0	68	20.4	1.65	(1.14-2.39)	
Dominant							
T/T	381	31.5	76	22.8	1.00		0.001832
T/G-G/G	830	68.5	257	77.2	1.55	(1.17-2.06)	
Recessive							
T/T-T/G	1005	83.0	265	79.6	1.00		0.1547
G/G	206	17.0	68	20.4	1.25	(0.92-1.70)	
Overdominant							
T/T-G/G	587	48.5	144	43.2	1.00		0.09005
T/G	624	51.5	189	56.8	1.23	(0.97-1.58)	
log-Additive							
0,1,2	1211	78.4	333	21.6	1.30	(1.09-1.55)	0.004112

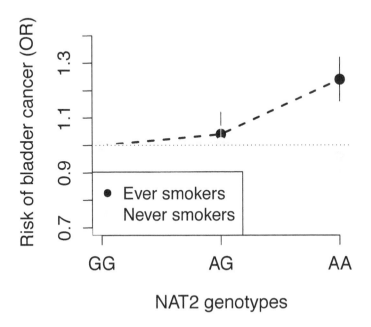

FIGURE 4.4
Interaction plot illustrating the different effect of NAT2 genotypes in never and ever smokers.

4.2.6 Gene × environment and gene × gene interactions

Gene-environment (G×E) interaction occurs when the effect of a given geno-type is different between different environmental exposures. Bladder cancer is an excellent example where polymorphisms in *NAT2* interact with smoking status to modulate bladder cancer risk [46, 127]. Figure 4.4 is an illustrative example of G×E interaction using the results found in [127]. The GG and AG genotypes, which tag rapid and intermediate acetylator, conferred a sig-nificantly increased bladder cancer risk (OR: 1.15; 95% CI: 1.09-1.22) with respect to the AA genotype, which tags the slow acetylator. However, the association is only evident in ever smokers (OR: 1.24; 95% CI: 1.16-1.32), but not in never smokers (OR: 0.96; 95% CI: 0.86-1.08), showing a significant genotypesmoking interaction. See Figure 4.4 where CIs do not overlap in the AA individuals.

G×E analyses can be performed within *SNPassoc*. Assume that we are interested in testing whether the risk of rs1422993 for asthma under the dominant model is different among smokers (variable smoke; 0=never, 1=ever). The `association` function fits a model with an interaction term where the environmental variable is a factor `factor`.

```
> association(casecontrol ~ dominant(rs1422993)*factor(smoke),
+                data=asthma.s)

      SNP: dominant(rs1422993  adjusted by:
 Interaction
---------------------
            0        OR lower upper   1        OR lower upper
G/G      486 130 1.00     NA      NA 242 43 0.66   0.46   0.97
G/T-T/T  354 128 1.35   1.02   1.79 150 38 0.95   0.63   1.42


p interaction: 0.8513

 factor(smoke) within dominant(rs1422993
---------------------
G/G
     0    1   OR lower upper
0 486 130 1.00    NA     NA
1 242   43 0.66  0.46   0.97


G/T-T/T
     0    1  OR lower upper
0 354  128 1.0    NA     NA
1 150   38 0.7   0.47   1.06


p trend: 0.8513

 dominant(rs1422993 within factor(smoke)
---------------------
0
            0    1   OR lower upper
G/G      486 130 1.00    NA     NA
G/T-T/T  354 128 1.35  1.02   1.79


1
            0  1   OR lower upper
G/G      242 43 1.00    NA     NA
G/T-T/T  150 38 1.43   0.88   2.31


p trend: 0.8513
```

The result is an interaction table showing that the risk of individuals carrying the T allele increases the risk of asthma in never smokers (OR=1.35; CI: 1.02-1.79) while it is not significant in ever smokers (OR=0.95; CI: 0.63-1.42). However, the interaction is not statistically significant (*P*-interaction=0.8513). The output also shows the stratified ORs that can help in interpreting the results.

In a similar way, G×G interactions of a given epistasis model can also be fitted using *SNPassoc*. In that case, the genetic model of the interacting SNP must be indicated in the `model.inteaction` argument.

```
> association(casecontrol ~ rs1422993*factor(rs184448),
+               data=asthma.s, model.interaction = "dominant" )

      SNP: rs1422993  adjusted by:
 Interaction
---------------------
          T/T      OR lower upper T/G      OR lower upper   0  1  G/G lower
G/G       227 43 1.00    NA    NA 359 96 1.41  0.95  2.10 128 30 1.24  0.74
G/T-T/T   154 33 1.13  0.69  1.86 265 93 1.85  1.24  2.77  78 38 2.57  1.55
          upper
G/G        2.07
G/T-T/T    4.27

p interaction: 0.24499

 factor(rs184448) within rs1422993
---------------------
G/G
          0  1   OR lower upper
T/T 227 43 1.00    NA    NA
T/G 359 96 1.41  0.95  2.10
G/G 128 30 1.24  0.74  2.07

G/T-T/T
          0  1   OR lower upper
T/T 154 33 1.00    NA    NA
T/G 265 93 1.64  1.05  2.55
G/G  78 38 2.27  1.32  3.90

p trend: 0.24499

 rs1422993 within factor(rs184448)
---------------------
T/T
          0  1   OR lower upper
G/G      227 43 1.00    NA    NA
G/T-T/T  154 33 1.13  0.69  1.86

T/G
          0  1   OR lower upper
G/G      359 96 1.00    NA    NA
G/T-T/T  265 93 1.31  0.95  1.82

G/G
          0  1   OR lower upper
G/G      128 30 1.00    NA    NA
G/T-T/T   78 38 2.08  1.19  3.62

p trend: 0.12743
```

We observe that the interaction between these two SNPs is not statistically significant ($P=0.24$). However, the OR of GG genotype of rs184448 differs across individuals between the GG and GT-TT genotypes of rs1422993 (see ORs for GG in the second table of the output).

4.3   Haplotype analysis

In genetic association studies, a haplotype is considered as a set of alleles across multiple loci that that occurs in a chromosome of a population. The presence of frequent haplotypes in genomic regions induces a correlation among SNPs, known as linkage disequilibrium (LD). The characterization of the correlational structure in the genomic SNP data of a population allows the reduction of the SNPs that need to be sequenced in a particular study. The genotyped SNPs can be used to impute the unobserved SNPs through the haplotypic structure. In addition, association studies that are based on genotyped SNPs may identify SNPs around a causal SNP that has not been genotyped, motivating analyses of higher resolution within the region. Therefore, the knowledge of haplotype structures across different populations is crucial for investigating the genetics of common diseases, as demonstrated by the International HapMap Project [25].

Genetic association studies can be extended from single SNP associations to haplotype associations[179]. Several examples including different complex diseases can be found in [180, 94, 100, 44, 5]. While alleles naturally occur in haplotypes, as they belong to chromosomes, the phase, or the knowledge of the chromosome an allele belongs to is lost in the genotyping process. As each genotype is measured with a different probe the phase between the alleles in a chromosome is broken. Consider for instance an individual for which one chromosome has alleles A and T at two different loci and alleles G or C at the second chromosome. The individual's genotypes for the individual at the two loci are A/G and T/C, which are the same genotypes of another individual that has alleles A and C in one chromosome and G and T in the second chromosome. Clearly, if the pair A-T confers a risk to a disease, subject one is at risk while subject two is not, despite both of them having the same genotypes.

The only unequivocal method of resolving phase ambiguity is sequencing the chromosomes of individuals. However, given the correlational structure of SNPs, it is possible to estimate the probability of a particular haplotype in a subject. This can be done in a genetic association study where a number of cases and controls are genotyped. There are numerous methods to infer unobserved haplotypes, two of the most popular are maximum likelihood, implemented via the expectation-maximization (EM) algorithm [34, 84], and a parsimony method [20]. Recent methods based on Bayesian models have also been proposed [146, 145]. In addition, haplotypes inferences carry uncertainty, which should be considered in association analyses.

4.3.1 Linkage disequilibrium heatmap plots

Linkage-disequilibrium (LD) is the non-random association of alleles in a chromosome. LD is a measure of the statistical association (e.g. correlation) between the allele frequencies at two different loci in a population. The correlation between neighboring SNPs, or the LD between their allele frequencies, depends on many factors such as recombination, mutation rates, selection, and genetic drift among many others. As such, the patterns between pair-wise SNP correlations describe the genetic history of a particular genomic region of the population. LD patterns in small genomic regions (<0.2Mb) are sustained by the distribution of specific haplotypes, which result in the non-random association of alleles. LD plots are therefore useful to visualize haplotype blocks and are implemented in the *LDheatmap* and *genetics*

```
> library(LDheatmap)
> library(genetics)
```

Haplotype visualization requires SNP annotation, with respect to a reference genome. The Bioconductor package *biomaRt* is a prime source of genomic annotation for the reference genomes of several species. A large amount of data is annotated for the reference genomes by the Ensembl genome database project (http://www.ensembl.org/index.html). Another source of annotation of reference genomes is at the genome browser (http://genome.ucsc.edu/). While data is available through web-applications and ftp servers, *biomaRt* allows the retrieval of data within R. The function useMart defines the reference genome and database from which data will be downloaded. listMarts() lists the available database that can be inquired and listDatasets lists the datasets available for the given database. We query the database ENSEMBL_MART_SNP and retrieve the hsapiens_snp.

```
> library(biomaRt)
> mart <- useMart("ENSEMBL_MART_SNP", dataset = "hsapiens_snp")
> nrow(listFilters(mart))
[1] 34
> head(listFilters(mart))
           name                description
1      chr_name Chromosome/scaffold name
2         start                      Start
3           end                        End
4    band_start                 Band Start
5      band_end                   Band End
6 marker_start               Marker Start
> listFilters(mart)[11,]
          name                                         description
11 snp_filter Filter by Variant mame (e.g. rs123, CM000001)
```

The function listFilters lists the query variables that are available for a given dataset. For instance for the dataset that encodes the annotation of human SNPs (hsapiens_snp) there are 34 query terms or filters, the 11th term is snp_filter that can be used to retrieve annotated information of the SNPs by name (e.g. rs123). For our asthma example, we retrieve the SNP reference ID, the genomic positions and the reference alleles on the human reference genome, given by the query variables refsnp_id, chr_name, chrom_start and allele

```
> snps <- labels(asthma.s)
> snpInfo <- getBM(c("refsnp_id", "chr_name", "chrom_start", "allele"),
+                  filters = c("snp_filter"),
+                  values = snps, mart = mart)
> head(snpInfo)
  refsnp_id chr_name chrom_start allele
1 rs4941643       13    49494251  G/A/T
2 rs3794381       13    49499427    C/G
3 rs2031532       13    49506711  A/C/G
4 rs2247119       13    49513006    T/C
5 rs8000149       13    49515889    C/T
6 rs2274276       13    49521817  C/A/G
```

We now want to observe the LD Heatmap plot of SNPs located in the region 34.5-35.0Mb of chromosome 7, for which we create a *snp* object with setupSNP that includes the selected SNPs

```
> mask <- with(snpInfo, chr_name=="7" & chrom_start>34.5e6 &
+                 chrom_start<35.0e6)
> snps.sel <- snpInfo[mask, "refsnp_id"]
> sel <- which(names(asthma)%in%snps.sel)
> asthma.hap <- setupSNP(asthma, sel, sep="")
```

We then create a *genotype* object from the *genetics* package

```
> snp.pos <- snpInfo[mask, "chrom_start"]
> snp.geno <- data.frame(lapply(asthma.hap[, snps.sel], genotype))
```

The LD plot follows from the calling plot on snp.geno and snp.pos

```
> LDplot <- LDheatmap(snp.geno, LDmeasure = "r",
+   title = "Pairwise LD in r^2", add.map = TRUE,
+   color = grey.colors(20), name = "myLDgrob", add.key = TRUE,
+   flip=TRUE, SNP.name=snps.sel)
```

The Figure 4.5 shows the pair-wise LD between the selected SNPs, measured by the squared of the correlation coefficient R^2 between the SNPs. In particular, we can see three SNP blocks. The package *LDheatmap* contains other features to improve the plots [137]. Note that SNPs are equidistant in the figure. They can be shown in their genomic positions by setting the argument genetic.distances = snp.pos in the function LDheatmap.

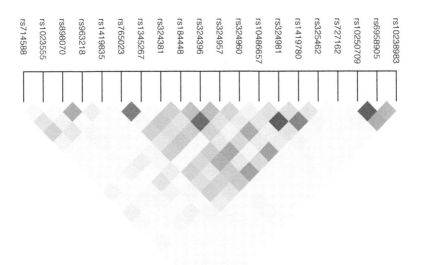

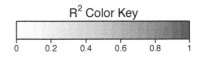

FIGURE 4.5
Linkage disequilibrium heatmap of selected SNPs located in chr7:34500000-35000000 region belonging to the asthma dataset.

4.3.2   Haplotype estimation

We now illustrate how to perform haplotype estimation from genotype data using the EM algorithm and how to integrate haplotype uncertainty when evaluating the association between traits and haplotypes. One of the SNP blocks in Figure 4.5 comprises rs714588, rs1023555 and rs898070. Haplotype inference is performed with *haplo.stats*, for which genotypes are encoded in a different format. `make.geno`, from *snpStats*, formats data for *haplo.stats*. The function `haplo.em` computes the haplotype frequency in the data for the SNPs of interest

```
> library(haplo.stats)
> snpsH <- c("rs714588", "rs1023555",  "rs898070")
> genoH <- make.geno(asthma.hap, snpsH)
> em <- haplo.em(genoH, locus.label = snpsH, miss.val = c(0, NA))
> em
==============================================================================
                                   Haplotypes
==============================================================================
   rs714588 rs1023555 rs898070 hap.freq
1         1         1        1  0.04090
2         1         1        2  0.02439
3         1         2        1  0.04264
4         1         2        2  0.44047
5         2         1        1  0.08271
6         2         1        2  0.08403
7         2         2        1  0.20794
8         2         2        2  0.07691
==============================================================================
                                    Details
==============================================================================
lnlike =  -4102.691
lr stat for no LD =  774.5411 , df =  4 , p-val =  0
```

Coding the common and variant alleles as 1 and 2, we can see there are 8 possible haplotypes across the subjects and are listed with an estimated haplotype frequency. Clearly, the haplotypes are not equally probable, as expected from the high LD between the SNPs. In particular, we observe that haplotypes 4 and 7 are the most probable, accumulating 65% of the haplotype sample. `haplo.em` estimates for each subject the probability of a given haplotype in each of the subject's chromosomes.

4.3.3   Haplotype association

We then want to assess if any of these haplotypes significantly associates with asthma. The `haplo.glm` fits a regression model between the phenotype and the haplotypes, incorporating the uncertainty for the probable haplotypes of individuals. The function `intervals` of *SNPassoc* provides a nice summary of the results

```
> trait <- asthma.hap$casecontrol
> mod <- haplo.glm(trait ~ genoH,
+                  family="binomial",
+                  locus.label=snpsH,
+                  allele.lev=attributes(genoH)$unique.alleles,
+                  control = haplo.glm.control(haplo.freq.min=0.05))
> intervals(mod)
                 freq   or   95%   C.I.   P-val
          ATG 0.4405   1.00 Reference haplotype
          GAA 0.0827   1.09 (  0.77 -   1.56 )   0.6160
          GAG 0.0841   0.99 (  0.69 -   1.42 )   0.9642
          GTA 0.2080   1.06 (  0.84 -   1.34 )   0.6379
          GTG 0.0769   1.09 (  0.76 -   1.58 )   0.6367
   genoH.rare 0.1079   1.15 (  0.83 -   1.59 )   0.3968
```

haplo.glm fits a logistic regression model for the asthma status (trait) on the inferred haplotypes (genoH), the names of SNPs and their allele names are passed in the locus.label and allele.lev arguments, while only haplotypes with at least 5% frequency are considered. As a result, we obtain the OR for each haplotype with their significance P-value, with respect to the most common haplotype (ATG with 44% frequency). In particular, from this analysis, we cannot see any haplotype significantly associated with asthma.

4.3.4   Sliding window approach

The inference of haplotypes depends on a predefined region or sets of SNPs. For instance, in the last section, we selected three SNPs that were in high LD. However, when no previous knowledge is available about the region or SNPs for which haplotypes should be inferred, we can apply a sliding window for haplotype inference [38, 178].

To illustrate this type of analysis, we now consider the second block of 10 SNPs in our asthma example (from 6th to 15th SNP in Figure 4.5). Considering large haplotypes, however, increases the number of possible haplotypes in the sample, decreasing the power of finding real associations. In addition, in predefined blocks, it is possible to miss the most efficient length of the susceptible haplotype, incrementing the loss of power. This is overcome using a sliding window. We thus ask which is the haplotype combination from any of 4, 5, 6 or 7 consecutive SNPs that gives the highest association with asthma status. We then reformat SNP genotypes in the region with the make.geno function and perform an association analysis for multiple haplotypes of i SNPs sliding from the 6th to the 15th SNP in the data. We perform an analysis for each window length i varying from 4 to 7 SNPs.

```
> snpsH2 <- colnames(snp.geno)[6:15]
> genoH2 <- make.geno(asthma.hap, snpsH2)
> haplo.score <- list()
> for (i in 4:7) {
+   trait <- asthma.hap$casecontrol
+   haplo.score[[i-3]] <- haplo.score.slide(trait, genoH2,
```

```
+                                trait.type="binomial",
+                                n.slide=i,
+                                simulate=TRUE,
+                                sim.control=score.sim.control(min.sim=100,
+                                             max.sim=200))
+ }
```

The results can be visualized with the following plot

```
> par(mfrow=c(2,2))
> for (i in 4:7) {
+     plot(haplo.score[[i-3]])
+     title(paste("Sliding Window=", i, sep=""))
+ }
```

In Figure 4.6, we observe that the highest $-\log_{10}(P)$-value is obtained for a haplotype of 4 SNP length starting at the 4th SNP of the selected SNPs. After deciding the best combination of SNPs, the haplotype association with asthma can be estimated by

```
> snpsH3 <- snpsH2[4:7]
> genoH3 <- make.geno(asthma.hap, snpsH3)
> mod <- haplo.glm(trait~genoH3,
+                    family="binomial",
+                    locus.label=snpsH3,
+                    allele.lev=attributes(genoH3)$unique.alleles,
+                    control = haplo.glm.control(haplo.freq.min=0.05))
> intervals(mod)
                 freq   or   95%   C.I.    P-val
      TCGT 0.3316   1.00 Reference haplotype
      GCAC 0.1463   1.53 (   1.17 -   2.00 )  0.0021
      GTAC 0.2826   1.28 (   1.01 -   1.60 )  0.0379
      TCGC 0.2272   1.16 (   0.91 -   1.48 )  0.2184
 genoH3.rare 0.0123 2.25 (   1.10 -   4.61 )  0.0261
```

Here, we observe that individuals carrying the haplotype GCAC have a 53% increased risk of asthma relative to those having the reference haplotype TCGT (OR=1.53, p=0.0021). The haplotype GTCA is also significantly associated with the disease (p=0.0379). A likelihood ratio test for haplotype status can be extracted from the results of the `haplo.glme` function:

```
> lrt <- mod$lrt
> pchisq(lrt$lrt, lrt$df, lower=FALSE)
[1] 0.008772248
```

We can also test the association between asthma and the haplotype adjusted for smoking status

```
> smoke <- asthma.hap$smoke
> mod.adj.ref <- glm(trait ~ smoke, family="binomial")
> mod.adj <- haplo.glm(trait ~ genoH3 + smoke ,
+                    family="binomial",
```

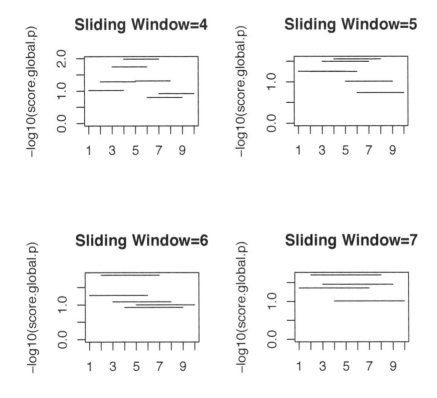

FIGURE 4.6
Sliding window approach varying haplotype size from 4 up to 7 SNPs located
in chr7:34500000-35000000 region from asthma data example

```
+                    locus.label=snpsH3,
+                    allele.lev=attributes(genoH3)$unique.alleles,
+                    control = haplo.glm.control(haplo.freq.min=0.05))
>
> lrt.adj <- mod.adj.ref$deviance - mod.adj$deviance
> pchisq(lrt.adj, mod.adj$lrt$df, lower=FALSE)
[1] 0.01460222
```

4.4   Genetic score

A genetic score, also called a polygenic risk score or genetic risk score, is given by the number of risk alleles that an individual carries [31]. Genetic scores aim to assess the collective prediction of the phenotype risk associated with multiple SNPs. The SNPs to be included in a genetic score can be selected from different sources. They may be selected from candidate genes, from reported SNPs associated with the trait, or from significant SNPs from a given analysis.

We illustrate how to perform a genetic risk score analysis with the asthma study, using the package *PredictABEL*. We start by performing the single SNP analysis, as previously discussed

```
> library(PredictABEL)
> data(asthma, package = "brgedata")
> dd.s <- setupSNP(asthma, 7:ncol(asthma), sep="")
> ans <- WGassociation(casecontrol, dd.s, model="log-add")
```

We choose the SNPs that show an association at 0.1 level to be included in a multivariate model,

```
> sel <- labels(dd.s)[additive(ans)<0.1]
> dd.sel <- dd.s[,sel]
> head(dd.sel)
  rs1422993 rs184448 rs324957 rs324960 rs324981 rs727162 rs6084432
1       G/G      T/T      G/G      C/T      A/T      G/G       G/G
2       G/T      G/G      A/C      C/C      A/A      G/G       G/G
3       G/G      T/T      G/G      C/T      T/T      G/C       G/A
4       G/T      T/G      G/A      C/C      A/T      G/G       G/G
5       G/T      T/T      G/G      C/C      A/T      G/G       G/G
6       G/G      T/T      G/G      C/T      A/T      G/C       G/G
```

We join the SNPs and case/control variable in a single *data.frame*. Using the `additive`, we recode the SNP genotypes as 0, 1 or 2 where the homozygous for the alternative allele is 2. As such, the genetic score is the sum of alternative alleles over selected SNPs

```
> dd.sel <- data.frame(lapply(dd.sel, additive))
> dd.end <- data.frame(casecontrol=dd.s$casecontrol, dd.sel)
```

```
> head(dd.end)
  casecontrol rs1422993 rs184448 rs324957 rs324960 rs324981 rs727162
1           0         0        0        0        1        1        0
2           0         1        2        2        0        0        0
3           0         0        0        0        1        2        1
4           0         1        1        1        0        1        0
5           1         1        0        0        0        1        0
6           0         0        0        0        1        1        1
  rs6084432
1         0
2         0
3         1
4         0
5         0
6         0
```

A multivariate regression model is fitted with all the SNPs to further select those that best predict the asthma status. An automatic variable selection, based on Akaike information criteria (AIC), can be applied to the model using the `stepAIC` function. Note that the function does not support missing values and requires having complete cases (i.e. no missing values on genotypes).

```
> dd.end.complete <- dd.end[complete.cases(dd.end),]
> mod <- stepAIC(glm(casecontrol ~ ., dd.end.complete,
+                family="binomial"),
+              method="forward", trace=0)
> snps.score <- names(coef(mod))[-1]
> snps.score
[1] "rs1422993" "rs324957"  "rs727162"  "rs6084432"
```

The selected SNPs, from the multivariate model, in rs1422993, rs324957, rs727162, rs6084432 are then used to create the genetic score. First, a summary shows the SNPs' associations of the selected SNPs with the trait

```
> summary(mod)

Call:
glm(formula = casecontrol ~ rs1422993 + rs324957 + rs727162 +
    rs6084432, family = "binomial", data = dd.end.complete)

Deviance Residuals:
    Min       1Q   Median       3Q      Max
-0.9775  -0.7290  -0.6655  -0.5596   1.9655

Coefficients:
            Estimate Std. Error z value Pr(>|z|)
(Intercept) -1.77498    0.13635 -13.018  < 2e-16 ***
rs1422993    0.20528    0.09849   2.084  0.03714 *
rs324957     0.25937    0.09266   2.799  0.00512 **
rs727162     0.18037    0.10447   1.727  0.08424 .
rs6084432    0.19982    0.11388   1.755  0.07932 .
---
Signif. codes:  0 '***' 0.001 '**' 0.01 '*' 0.05 '.' 0.1 ' ' 1
```

```
(Dispersion parameter for binomial family taken to be 1)

    Null deviance: 1587.0  on 1517  degrees of freedom
Residual deviance: 1569.2  on 1513  degrees of freedom
AIC: 1579.2

Number of Fisher Scoring iterations: 4
```

The function **riskScore** function of the *PredictABEL* package creates the risk score

```
> pos <- which(names(dd.end.complete)%in%snps.score)
> names(dd.end.complete)
[1] "casecontrol" "rs1422993"   "rs184448"    "rs324957"    "rs324960"
[6] "rs324981"    "rs727162"    "rs6084432"
> pos
[1] 2 4 7 8
> score <- riskScore(mod, data=dd.end.complete,
+                     cGenPreds=pos,
+                     Type="unweighted")
> table(score)
score
  0   1   2   3   4   5   6
101 389 491 347 139  44   7
```

The distribution of the score across individuals can be seen in Figure 4.7

```
> hist(score, col="gray90")
```

Once the genetic score is created, we test its association with asthma status using a general linear model

```
> mod.lin <- glm(casecontrol~score, dd.end.complete,
+                family="binomial")
> exp(coef(mod.lin)[2])
   score
1.240039
```

We observe that the risk of asthma increases 21% per each risk allele. The predictive power of the genetic score can be assessed by computing the area under the ROC curve (AUC)

```
> predrisk <- predRisk(mod.lin, dd.end.complete)
> plotROC(data=dd.end.complete, cOutcome=1,
+         predrisk = predrisk)
```

```
AUC [95% CI] for the model 1 :  0.574 [ 0.54  -  0.607 ]
```

Figure 4.8 shows that the predictive power of the genetic score is 57.4% with a 95% confidence interval (54.0 - 60.7).

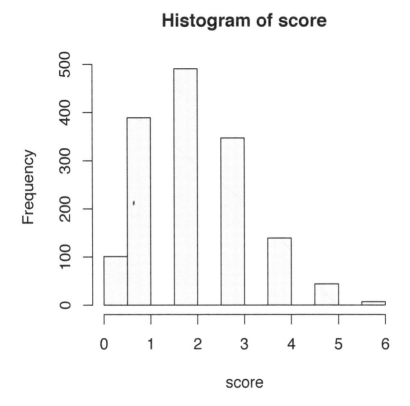

FIGURE 4.7

Distribution of the genetic score used to predict case/control status in the asthma example.

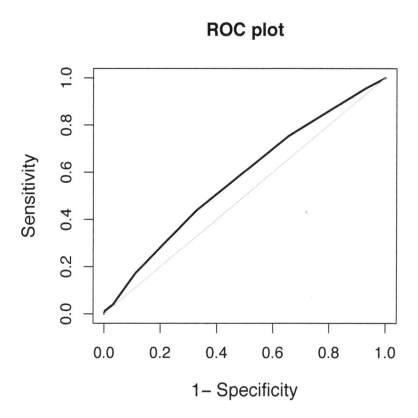

FIGURE 4.8
ROC curve of the genetic score used to predict case/control status in the
asthma example.

4.5   Genome-wide association studies

Genome-wide association studies (GWASs) assess the association between the trait of interest and up to millions of SNPs. GWASs have been used to discover thousands of SNPs associated with several complex diseases [87]. The basic statistical methods are similar to those previously described, in particular, the massive univariate testing. The main issue with GWASs is data management and computation. Most publicly available data is in PLINK format, where genomic data is stored in a binary BED file, and phenotype and annotation data in text BIM and FAM files. PLINK data can be loaded into R with the Bioconductor's package *snpStats* (see Section 3.2).

We illustrate the analysis of a GWAS including 100,000 SNPs that have been simulated using real data from a case-control study. Our phenotype of interest is obesity (0: not obese; 1: obese) that has been created using body mass index information of each individual. We start by loading genotype data that are in PLINK format (`obesity.bed`, `obesity.bim`, `obesity.fam` files) in the *brgedata* package

```
> library(snpStats)
> path <- system.file("extdata", package="brgedata")
> ob.plink <- read.plink(file.path(path, "obesity"))
```

The imported object is a list containing the genotypes, the family structure and the SNP annotation.

```
> names(ob.plink)
[1] "genotypes" "fam"       "map"
```

We store genotype, annotation and family data in different variables for downstream analyses

```
> ob.geno <- ob.plink$genotypes
> ob.geno
A SnpMatrix with  2312 rows and  100000 columns
Row names:  100 ... 998
Col names:  MitoC3993T ... rs28600179
>
> annotation <- ob.plink$map
> head(annotation)
          chromosome  snp.name cM position allele.1 allele.2
MitoC3993T         NA MitoC3993T NA     3993        T        C
MitoG4821A         NA MitoG4821A NA     4821        A        G
MitoG6027A         NA MitoG6027A NA     6027        A        G
MitoT6153C         NA MitoT6153C NA     6153        C        T
MitoC7275T         NA MitoC7275T NA     7275        T        C
MitoT9699C         NA MitoT9699C NA     9699        C        T
>
> family <- ob.plink$fam
> head(family)
```

```
        pedigree member father mother sex affected
100      FAM_OB     100    NA     NA   1        1
1001     FAM_OB    1001    NA     NA   1        1
1004     FAM_OB    1004    NA     NA   2        2
1005     FAM_OB    1005    NA     NA   1        2
1006     FAM_OB    1006    NA     NA   2        1
1008     FAM_OB    1008    NA     NA   1        1
```

Notice that **geno** is an object of class *SnpMatrix* that stores the SNPs in binary (raw) format. While some basic phenotype data is usually available in the **fam** field of the *SnpMatrix* object, a more complete phenotypic characterization of the sample is usually distributed in additional text files. In our example, the complete phenotype data is in a tab-delimited file

```
> ob.pheno <- read.delim(file.path(path, "obesity.txt"))
> head(ob.pheno)
    id gender obese age   smoke country
1 4180   Male     1  41 Current      50
2 4880 Female    NA  35      Ex      51
3  435   Male     1  50      Ex      53
4 4938   Male     0  44 Current      53
5 2977   Male    NA  49   Never      53
6 1705   Male     0  40   Never      50
```

The file contains phenotypic information for a different set of individuals that overlap with those in the **ob.geno** object. Therefore, before analysis, we need to correctly merge and order the individuals across genomic and phenotype datasets. The row names of **ob.geno** correspond to the individual identifiers (id) variable of **ob.pheno**. Consequently, we also rename the rows of **ob.pheno** with the **id** variable

```
> rownames(ob.pheno) <- ob.pheno$id
```

We can check if the row names of the datasets match

```
> identical(rownames(ob.pheno), rownames(ob.geno))
[1] FALSE
```

FALSE indicates that either there are different individuals in both objects or that they are in different order. This can be fixed by selecting common individuals.

```
> ids <- intersect(rownames(ob.pheno), rownames(ob.geno))
> geno <- ob.geno[ids, ]
> ob <- ob.pheno[ids, ]
> identical(rownames(ob), rownames(geno))
[1] TRUE
>
> family <- family[ids, ]
```

4.5.1 Quality control of SNPs

We now perform the quality control (QC) of genomic data at the SNP and individual levels, before association testing [3]. Different measures can be used to perform QC and remove: 1) SNPs with a high rate of missing; 2) rare SNPS (e.g. having low minor allele frequency - MAF); and 3) SNPs that do not pass the HWE test.

Typically, markers with a call rate less than 95% are removed from association analyses, although some large studies chose higher call-rate thresholds (99%). Markers of low MAF ($<5\%$) are also filtered. The significance threshold rejecting a SNP for not being in HWE has varied greatly between studies, from thresholds between 0.001 and 5.7×10^{-7} [21]). Including SNPs with extremely low P-values for the HWE test will require individual examination of the SNP genotyping process. A parsimonious threshold of 0.001 may be considered, though robustly genotyped SNPs below this threshold may remain in the study [3], as deviations from HWE may indeed arise from biological processes.

The function `col.summary` offers different summaries (at SNP level) that can be used in QC

```
> info.snps <- col.summary(geno)
> head(info.snps)
            Calls Call.rate Certain.calls      RAF        MAF
MitoC3993T  2286 0.9887543             1 0.9851269 0.0148731409
MitoG4821A  2282 0.9870242             1 0.9982472 0.0017528484
MitoG6027A  2307 0.9978374             1 0.9956654 0.0043346337
MitoT6153C  2308 0.9982699             1 0.9893847 0.0106152513
MitoC7275T  2309 0.9987024             1 0.9991338 0.0008661758
MitoT9699C  2302 0.9956747             1 0.9268028 0.0731972198
                  P.AA          P.AB      P.BB      z.HWE
MitoC3993T  0.0148731409 0.0000000000 0.9851269 -47.81213
MitoG4821A  0.0017528484 0.0000000000 0.9982472 -47.77028
MitoG6027A  0.0043346337 0.0000000000 0.9956654 -48.03124
MitoT6153C  0.0103986135 0.0004332756 0.9891681 -47.05069
MitoC7275T  0.0008661758 0.0000000000 0.9991338 -48.05206
MitoT9699C  0.0729800174 0.0004344049 0.9265856 -47.82555
```

snpStats does not compute P-values of the HWE test but computes its z-scores. A P-value of 0.001 corresponds to a z-score of ± 3.3 for a two-tail test. Strictly speaking, the HWE test should be applied to controls only (e.g. `obese = 0`), however, the default computation is for all samples.

We thus filter SNPs with a call rate $> 95\%$, MAF of $> 5\%$ and $z.HWE < 3.3$ in controls

```
> controls <- ob$obese ==0 & !is.na(ob$obese)
> geno.controls <- geno[controls,]
> info.controls <- col.summary(geno.controls)
>
> use <- info.snps$Call.rate > 0.95 &
+        info.snps$MAF > 0.05 &
+        abs(info.controls$z.HWE < 3.3)
```

```
> mask.snps <- use & !is.na(use)
>
> geno.qc.snps <- geno[ , mask.snps]
> geno.qc.snps
A SnpMatrix with  2312 rows and  88723 columns
Row names:  4180 ... 277
Col names:  MitoT9699C ... rs28562204
>
> annotation <- annotation[mask.snps, ]
```

It is common practice to report the number of SNPs that have been removed from the association analyses

```
> # number of SNPs removed for bad call rate
> sum(info.snps$Call.rate < 0.95)
[1] 888
> # number of SNPs removed for low MAF
> sum(info.snps$MAF < 0.05, na.rm=TRUE)
[1] 10461
> #number of SNPs that do not pass HWE test
> sum(abs(info.controls$z.HWE > 3.3), na.rm=TRUE)
[1] 80
> # The total number of SNPs do not pass QC
> sum(!mask.snps)
[1] 11277
```

4.5.2   Quality control of individuals

QC of individuals, or biological samples, comprises four main steps: 1) The identification of individuals with discordant reported and genomic sex, 2) the identification of individuals with outlying missing genotype or heterozygosity rate, 3) the identification of duplicated or related individuals, and 4) the identification of individuals of divergent ancestry from the sample [3].

We start by removing individuals with sex discrepancies, a large number of missing genotypes and outlying heterozygosity. The function `row.summary` returns the call rate and the proportion of called SNPs which are heterozygous per individual.

```
> info.indv <- row.summary(geno.qc.snps)
> head(info.indv)
     Call.rate Certain.calls Heterozygosity
4180 0.9998873             1      0.3426781
4880 0.9998197             1      0.3539180
435  0.9958297             1      0.3392188
4938 0.9994928             1      0.3411782
2977 0.9985348             1      0.3426004
1705 0.9936657             1      0.3357721
```

Gender is usually inferred from the heterozygosity of chromosome X. Males have an expected heterozygosity of 0 and females of 0.30. Chromosome X heterozygosity can be extracted using `row.summary` function and then plotted

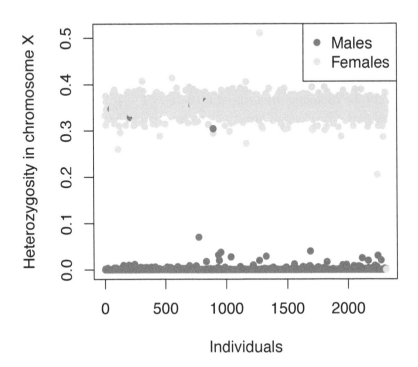

FIGURE 4.9

Heterozygosity in chromosome X by gender provided in the phenotypic data.

```
> geno.X <- geno.qc.snps[,annotation$chromosome=="23" &
+                        !is.na(annotation$chromosome)]
> info.X <- row.summary(geno.X)
> mycol <- ifelse(ob$gender=="Male", "gray40", "gray80")
> plot(info.X$Heterozygosity, col=mycol,
+       pch=16, xlab="Individuals",
+       ylab="Heterozygosity in chromosome X")
> legend("topright", c("Males", "Females"), col=mycol,
+        pch=16)
```

Figure 4.9 shows that there are some reported males with non-zero X-heterozygosity and females with zero X-heterozygosity. These samples are located in `sex.discrep` for later removal

```
> sex.discrep <- (ob$gender=="Male" & info.X$Heterozygosity > 0.2) |
+                 (ob$gender=="Female" & info.X$Heterozygosity < 0.2)
```

Sex filtering based on X-heterozygosity is not sufficient to identify rare aneuploidies, like XXY in males. Alternatively, plots of the mean allelic intensities of SNPs on the X and Y chromosomes can identify mis-annotated sex as well as sex chromosome aneuploidies.

Now, we identify individuals with outlying heterozygosity from the overall genomic heterozygosity rate that is computed by `row.summary`. Heterozygosity, can also be computed from the statistic $F = 1 - \frac{f(Aa)}{E(f(Aa))}$, where $f(Aa)$ is the observed proportion of heterozygous genotypes (Aa) of a given individual and $E(f(Aa))$ is the expected proportion of heterozygous genotypes. A subject's $E(f(Aa))$ can be computed from the MAF across all the subjects' non-missing SNPs

```
> MAF <- col.summary(geno.qc.snps)$MAF
> callmatrix <- !is.na(geno.qc.snps)
> hetExp <- callmatrix %*% (2*MAF*(1-MAF))
> hetObs <- with(info.indv, Heterozygosity*(ncol(geno.qc.snps))*Call.rate)
> info.indv$hetF <- 1-(hetObs/hetExp)
>
> head(info.indv)
     Call.rate Certain.calls Heterozygosity         hetF
4180 0.9998873             1      0.3426781  0.023324353
4880 0.9998197             1      0.3539180 -0.008701237
435  0.9958297             1      0.3392188  0.033025203
4938 0.9994928             1      0.3411782  0.027596273
2977 0.9985348             1      0.3426004  0.023487306
1705 0.9936657             1      0.3357721  0.042762824
```

In Figure 4.10, we compare F statistic and the `Heterozygosity` field obtained from `row.summary`

Individuals whose F statistic is outside the band ± 0.1 are considered sample outliers (left panel Figure 4.10) and correspond to those having a heterozygosity rate lower than 0.32.

GWASs are studies that are typically based on population samples. Therefore, close familial relatedness between individuals is not representative of the sample. We, therefore, remove individuals whose relatedness is higher than expected. The package *SNPRelate* is used to perform identity-by-descent (IBD) analysis, computing kinship within the sample. The package requires a data in a GDS format that is obtained with the function `snpgdsBED2GDS`. In addition, IBD analysis requires SNPs that are not in LD (uncorrelated). The function `snpgdsLDpruning` iteratively removes adjacent SNPs that exceed an LD threshold in a sliding window

```
> library(SNPRelate)
>
> # Transform PLINK data into GDS format
> snpgdsBED2GDS("obesity.bed",
```

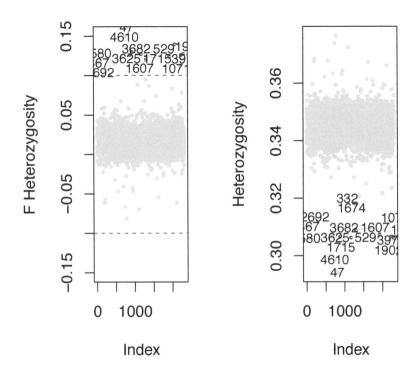

FIGURE 4.10
Heterozygosity computed using F statistic (left panel) and using row.summary function (right panel). The horizontal dashed line shows a suggestive value to detect individuals with outlier heterozygosity values.

```
+                "obesity.fam",
+                "obesity.bim",
+                out="obGDS")
Start snpgdsBED2GDS ...
BED file: "obesity.bed" in the SNP-major mode (Sample X SNP)
FAM file: "obesity.fam", DONE.
BIM file: "obesity.bim", DONE.
Wed May 15 16:12:38 2019  store sample id, snp id, position, and chromosome.
start writing: 2312 samples, 100000 SNPs ...
  Wed May 15 16:12:38 2019 0%
  Wed May 15 16:12:40 2019 100%
Wed May 15 16:12:40 2019  Done.
Optimize the access efficiency ...
Clean up the fragments of GDS file:
    open the file 'obGDS' (55.7M)
    # of fragments: 39
    save to 'obGDS.tmp'
    rename 'obGDS.tmp' (55.7M, reduced: 252B)
    # of fragments: 18
> genofile <- snpgdsOpen("obGDS")
>
> #Prune SNPs for IBD analysis
> set.seed(12345)
> snps.qc <- colnames(geno.qc.snps)
> snp.prune <- snpgdsLDpruning(genofile, ld.threshold = 0.2,
+                              snp.id = snps.qc)
SNP pruning based on LD:
Excluding 13,410 SNPs (non-autosomes or non-selection)
Excluding 0 SNP (monomorphic: TRUE, MAF: NaN, missing rate: NaN)
Working space: 2,312 samples, 86,590 SNPs
    using 1 (CPU) core
    sliding window: 500,000 basepairs, Inf SNPs
    |LD| threshold: 0.2
    method: composite
Chromosome 1: 31.51%, 2,433/7,721
Chromosome 2: 30.04%, 2,418/8,050
Chromosome 3: 30.84%, 2,059/6,676
Chromosome 4: 31.13%, 1,845/5,927
Chromosome 5: 30.87%, 1,875/6,074
Chromosome 6: 28.19%, 1,903/6,750
Chromosome 7: 31.24%, 1,673/5,356
Chromosome 8: 28.82%, 1,606/5,572
Chromosome 9: 31.52%, 1,487/4,718
Chromosome 10: 30.63%, 1,590/5,191
Chromosome 11: 31.12%, 1,485/4,772
Chromosome 12: 31.53%, 1,531/4,855
Chromosome 13: 31.19%, 1,136/3,642
Chromosome 14: 32.59%, 1,059/3,249
Chromosome 15: 31.80%, 973/3,060
Chromosome 16: 35.65%, 1,060/2,973
Chromosome 17: 36.97%, 1,006/2,721
Chromosome 18: 34.22%, 1,008/2,946
Chromosome 19: 40.80%, 694/1,701
Chromosome 20: 35.87%, 864/2,409
Chromosome 21: 34.62%, 485/1,401
Chromosome 22: 34.61%, 552/1,595
```

```
30,742 markers are selected in total.
> snps.ibd <- unlist(snp.prune, use.names=FALSE)
```

Note that this process is performed with SNPs that passed previous QC checks. IBD coefficients are then computed using the method of moments, implemented in **snpgdsIBDMoM**. The result of the analysis is a table indicating kinship among pairs of individuals

```
> ibd <- snpgdsIBDMoM(genofile, kinship=TRUE,
+                     snp.id = snps.ibd,
+                     num.thread = 1)
IBD analysis (PLINK method of moment) on genotypes:
Excluding 69,258 SNPs (non-autosomes or non-selection)
Excluding 0 SNP (monomorphic: TRUE, MAF: NaN, missing rate: NaN)
Working space: 2,312 samples, 30,742 SNPs
    using 1 (CPU) core
PLINK IBD:   the sum of all selected genotypes (0,1,2) = 32844904
Wed May 15 16:12:49 2019    (internal increment: 6656)

[..................................................] 0%, ETC: ---
[==================================================] 100%, completed in 14s
Wed May 15 16:13:04 2019    Done.
> ibd.kin <- snpgdsIBDSelection(ibd)
> head(ibd.kin)
  ID1  ID2        k0         k1     kinship
1 100 1001 0.9926611 0.00191261 0.003191305
2 100 1004 1.0000000 0.00000000 0.000000000
3 100 1005 1.0000000 0.00000000 0.000000000
4 100 1006 1.0000000 0.00000000 0.000000000
5 100 1008 1.0000000 0.00000000 0.000000000
6 100 1013 1.0000000 0.00000000 0.000000000
```

A pair of individuals with higher than expected relatedness are considered with kinship score > 0.1

```
> ibd.kin.thres <- subset(ibd.kin, kinship > 0.1)
> head(ibd.kin.thres)
          ID1  ID2        k0        k1    kinship
46484    1049  188 0.2731024 0.5431008 0.2276736
232848   1202 1330 0.0000000 0.0000000 0.5000000
281069   1237  872 0.2747742 0.4556623 0.2486973
640474    155 1682 0.2410303 0.4506688 0.2668176
806337    170 2015 0.2548016 0.5399619 0.2376087
1158509  2055  825 0.0000000 0.0000000 0.5000000
```

The ids of the individuals with unusual kinship are located with **related** function from the **SNPassoc** package

```
> ids.rel <-  related(ibd.kin.thres)
> ids.rel
 [1] "4364" "3380" "2999" "2697" "2611" "2088" "1202" "872"  "825"  "684"
[11] "188"  "170"  "155"  "2071"
```

Summing up, individuals with more than 3-7% missing genotypes [27, 139],

with sex discrepancies, F absolute value > 1 and kinship coefficient > 0.1 are
removed from the genotype and phenotype data

```
> use <- info.indv$Call.rate > 0.95 &
+       abs(info.indv$hetF) < 0.1 &
+       !sex.discrep &
+       !rownames(info.indv)%in%ids.rel
> mask.indiv <- use & !is.na(use)
> geno.qc <- geno.qc.snps[mask.indiv, ]
>
> ob.qc <- ob.pheno[mask.indiv, ]
> identical(rownames(ob.qc), rownames(geno.qc))
[1] TRUE
```

These QC measures are usually reported

```
> # number of individuals removed to bad call rate
> sum(info.indv$Call.rate < 0.95)
[1] 34
> # number of individuals removed for heterozygosity problems
> sum(abs(info.indv$hetF) > 0.1)
[1] 15
> # number of individuals removed for sex discrepancies
> sum(sex.discrep)
[1] 8
> # number of individuals removed to be related with others
> length(ids.rel)
[1] 14
> # The total number of individuals that do not pass QC
> sum(!mask.indiv)
[1] 70
```

4.5.3   Population ancestry

As GWAS are based on general population samples, individual genetic differences between individuals need to be also representative of the population at
large. The main source of genetic differences between individuals is ancestry.
Therefore, it is important to check that there are not individuals with unexpected genetic differences in the sample. Ancestral differences can be inferred
with principal component analysis (PCA) on the genomic data. Individuals
with outlying ancestry can be removed from the study while smaller differences in ancestry can be adjusted in the association models, including the first
principal components as covariates.

PCA on genomic data can be computed using the *SNPRelate* package
with the snpgdsPCA function. Efficiency can be improved by removing SNPs
that are in LD before PCA, see (snps.ibd) in the previous IBD analysis.
In addition snpgdsPCA allows parallelization with the argument num.thread
that determines the number of computing cores to be used

```
> pca <- snpgdsPCA(genofile, sample.id = rownames(geno.qc),
+                             snp.id = snps.ibd,
+                             num.thread=1)
Principal Component Analysis (PCA) on genotypes:
Excluding 69,258 SNPs (non-autosomes or non-selection)
Excluding 0 SNP (monomorphic: TRUE, MAF: NaN, missing rate: NaN)
Working space: 2,242 samples, 30,742 SNPs
    using 1 (CPU) core
PCA:    the sum of all selected genotypes (0,1,2) = 31854924
CPU capabilities: Double-Precision SSE2
Wed May 15 16:14:10 2019    (internal increment: 216)

[.....................................................]  0%, ETC: ---
[=====================================================] 100%, completed in 59s
Wed May 15 16:15:09 2019    Begin (eigenvalues and eigenvectors)
Wed May 15 16:15:16 2019    Done.
```

A PCA plot for the first two components can be obtained with

```
> with(pca, plot(eigenvect[,1], eigenvect[,2],
+                xlab="1st Principal Component",
+                ylab="2nd Principal Component",
+                main = "Ancestry Plot",
+                pch=21, bg="gray90", cex=0.8))
```

Inspection of Figure 4.11 can be used to identify individuals with unusual ancestry and remove them. Individuals with outlying values in the principal components will be considered for QC. In our example, we can see outlying individuals on the right side of the plot with 1st PC > 0.05. Smaller differences in ancestry are an important source of bias in association tests, as explained later. Therefore, we keep the first five principal components and add it to the phenotypic information that will be used in the association analyses

```
> ob.qc <- data.frame(ob.qc, pca$eigenvect[, 1:5])
```

After performing QC, the GDS file can be closed

```
> closefn.gds(genofile)
```

4.5.4 Genome-wide association analysis

Genome-wide association analysis involves regressing each SNP separately on our trait of interest. The analyses should be adjusted for clinical, environmental, and/or demographic factors as well as ancestral differences between the subjects. The analysis can be performed with a range of functions in *snpStats* package. We first examine the unadjusted whole genome association of our obesity study

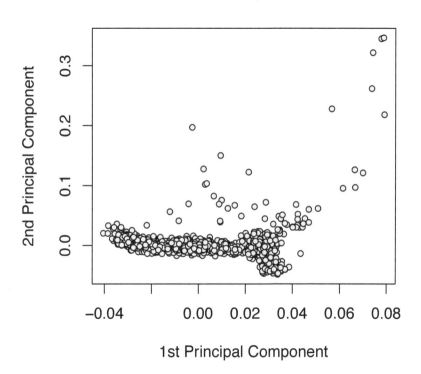

FIGURE 4.11
1st and 2nd principal components of obesity GWAS data example.

```
> res <- single.snp.tests(obese, data=ob.qc,
+                          snp.data=geno.qc)
> res[1:5,]
                N Chi.squared.1.df Chi.squared.2.df     P.1df      P.2df
MitoT9699C   2134        3.0263311               NA 0.08192307        NA
MitoA11252G  2090        0.3561812               NA 0.55063478        NA
MitoA12309G  2136        0.1776464               NA 0.67340371        NA
MitoG16130A  2069        2.4766387        3.9480296 0.11554896 0.1388981
rs28705211   2125        0.7277258        0.7546827 0.39362135 0.6856820
```

This analysis is only available for the additive ($\chi^2(1.\text{df})$) and the codominant models ($\chi^2(2.\text{df})$). It requires the name variable phenotype (**obese**) in the **data** argument. Genomic data are given in the **snp.data** argument. It is important that the individuals in the rows of both datasets match. SNPs in the mitochondrial genome and gonosomes return NA for the χ^2 estimates. These variants should be analyzed separately. A common interest is to analyze autosomes only, and therefore these SNPs can be removed in the QC process.

A quantitative trait can also be analyzed setting the argument **family** equal to **Gaussian**

```
> res.quant <- snp.rhs.tests(age ~ 1,  data=ob.qc,
+                            snp.data=geno.qc,
+                            family="Gaussian")
> head(res.quant)
            Chi.squared Df  p.value
MitoT9699C  0.003422591  1 0.953348
```

4.5.5 Adjusting for population stratification

Population stratification inflates the estimates of the χ^2 tests of association between the phenotype and the SNPs, and as a consequence the false positive rate increases. Figure 4.5.5 illustrates why population stratification may lead to false associations. In the hypothetical study in the figure, we compare 20 cases and 20 controls where individuals carrying a susceptibility allele are denoted by a dot. The overall frequency of the susceptibility allele is much larger in cases ($0.55 = 11/20$) than in controls ($0.35 = 7/20$), the odds of being a case in allele carriers is ~ 2.3 times higher than the odds of being a case in non carriers (OR= $2.27 = (0.55/0.45) / (0.35/0.65)$). However, the significant increase in susceptibility between the allele is misleading, as the OR in population A (light color) is 0.89 and in population B (dark color) is 1.08. The susceptibility allele strongly discriminates population A from B, and given the differences of the trait frequency between populations, it is likely that the association of the allele with the trait is through its links with population differences and not with the trait itself.

In genome-wide analyses, the inflation of the associations due to undetected latent variables is assessed by quantile-quantile (Q-Q) plots where observed χ^2 values are plotted against the expected ones

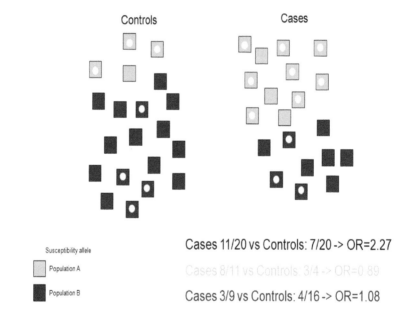

FIGURE 4.12
Illustrative example of population stratification. Read Section 4.2 for a detailed description.

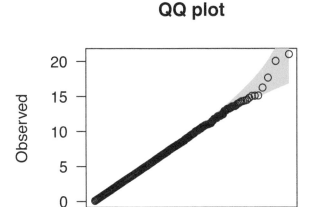

FIGURE 4.13
QQ-plot corresponding to obesity GWAS data example.

```
> chi2 <- chi.squared(res, df=1)
> qq.chisq(chi2)


           N       omitted        lambda
88723.000000    0.000000      1.003853
```

Figure 4.13 shows, in particular, that the χ^2 estimates are not inflated (λ is also close to 1), as all quantile values fall in the confidence bands, meaning that most SNPs are not associated with obesity. In addition, the Figure does not show any top SNP outside the confidence bands. A Q-Q plot with top SNPs outside the confidence bands indicates that those SNPs are truly associated with the disease and, hence, do not follow the null hypothesis. Therefore, the Q-Q plot of our examples reveals no significant SNP associations.

Q-Q plots are used to inspect population stratification. In particular, when population stratification is present, most SNP Q-Q values will be found outside the confidence bands, suggesting that the overall genetic structure of the sample can discriminate differences between subject traits. The λ value is a measure of the degree of inflation. The main source of population stratification that is derived from genomic data is ancestry. Therefore, in the cases of inflated Q-Q plots, it is ancestry differences and not individual SNP differences that explain the differences in the phenotype. Population stratification may be corrected by genomic control, mixed models or EIGENSTRAT method [114]. However, the most common approach is to use the inferred ancestry from genomic data as covariates in the association analyses [113]. Genome-wide association analysis typically adjusts for population stratification using the PCs on genomic data to infer ancestral differences in the sample. Covariates are easily incorporated in the model of `snp.rhs.tests`

```
> res.adj <- snp.rhs.tests(obese ~ X1 + X2 + X3 + X4 + X5,
+                          data=ob.qc, snp.data=geno.qc)
> head(res.adj)
           Chi.squared Df   p.value
MitoT9699C    3.108601  1 0.07787985
```

This function only computes the additive model, adjusting for the first five genomic PCs. The resulting $-\log_{10}(P)$-values of association for each SNP are then extracted

```
> pval.log10 <- -log10(p.value(res.adj))
```

These transformed P-values are used to create a Manhattan plot to visualize which SNPs are significantly associated with obesity. We use our function **manhattanPlot** as described in Section 2.3, although *qqman* package can also be used.

```
> library(tidyverse)
> library(ggplot2)
> library(ggrepel)
> # Create the required data frame
> pvals <- data.frame(SNP=annotation$snp.name,
+                     CHR=annotation$chromosome,
+                     BP=annotation$position,
+                     P=p.value(res.adj))
> # missing data is not allowed
> pvals <- subset(pvals, !is.na(CHR) & !is.na(P))
>
> manhattanPlot(pvals, color=c("gray90", "gray40"))
```

Significance at Bonferroni level is set at $10^{-7} = 0.05/10^5$, as we are testing 100,000 SNPs. The level corresponds to $-\log_{10}(P) = 6.30$. Therefore, we confirm, as expected form the Q-Q plot, that no SNP in our study is significantly associated with obesity, as observed in Figure 4.14. It should be notice that the standard Bonferroni significant level in GWASs is considered as 5×10^{-8} since SNParray data use to contain 500K-1M SNPs [108].

With our obesity example, we illustrate the common situation of finding no significant associations in small studies (thousands of subjects) with small genomic data (100,000 SNPs). This situation motivates multi-center studies with larger samples sizes, where small effects can be inferred with sufficient power and consistency.

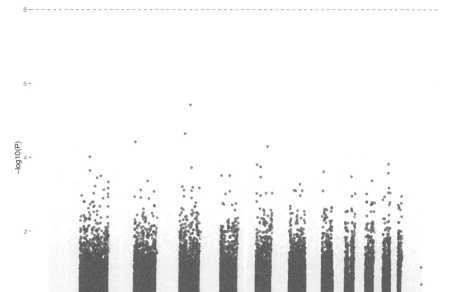

FIGURE 4.14
Manhattan plot of obesity GWAS data example.

The *snpStats* package performs association analyses using either codominant or additive models and only provides their p-values. Let us imagine we are interested in using *SNPassoc* to create the tables with the different genetic models for the most significant SNPs. This can be performed as following. We first select the SNPs that pass a p-value threshold. For instance, 10^{-5} (in real situations it should be the GWAS level 5×10^{-8})

```
> topPvals <- subset(pvals, P<10e-5)
> topSNPs <- as.character(topPvals$SNP)
```

We then export the data into a text file that can be imported into R as a `data.frame` that can be analyzed using *SNPassoc*

```
> # subset top SNPs
> geno.topSNPs <- geno.qc[, topSNPs]
> geno.topSNPs
A SnpMatrix with  2242 rows and  7 columns
Row names:  4180 ... 277
Col names:  rs2769689 ... rs285729
> # export top SNPs
> write.SnpMatrix(geno.topSNPs, file="topSNPs.txt")
[1] 2242    7
> # import top SNPs
> ob.top <- read.delim("topSNPs.txt", sep="")
> # add phenotypic information(ids are in the same order)
> ob.top <- cbind(ob.top, ob.qc)
> # prepare data for SNPassoc (SNPs are coded as 0,1,2)
> ii <- grep("^rs", names(ob.top))
> ob.top.s <- setupSNP(ob.top, colSNPs = ii,
+                      name.genotypes=c(0,1,2))
> # run association (all)
> WGassociation(obese, ob.top.s)
            comments codominant dominant recessive overdominant
rs2769689      -      0.00059   0.00353   0.00097     0.08474
rs10193241     -      0.00026   0.19355   0.00005     0.00044
rs7349742      -      0.00009   0.01839   0.00004     0.00182
rs12201995     -      0.00006   0.00185   0.00011     0.01271
rs6931936      -      0.00001   0.00078   0.00002     0.00368
rs1935960      -      0.00006   0.74717   0.00001     0.00001
rs285729       -      0.00027   0.01271   0.00022     0.00335
            log-additive
rs2769689      0.00013
rs10193241     0.00013
rs7349742      0.00002
rs12201995     0.00001
rs6931936      0.00000
rs1935960      0.00009
rs285729       0.00006
> # run association (one)
> association(obese ~ rs2769689, ob.top.s)

SNP: rs2769689  adjusted by:
                0    %   1    %   OR lower upper   p-value   AIC
Codominant
A/A           115  6.6  44 11.1 1.00               0.0005938 2042
```

```
A/B            663 38.1 170 42.8 0.67  0.46  0.99
B/B            961 55.3 183 46.1 0.50  0.34  0.73
Dominant
A/A            115  6.6  44 11.1 1.00                0.0035270 2047
A/B-B/B       1624 93.4 353 88.9 0.57  0.39  0.82
Recessive
A/A-A/B        778 44.7 214 53.9 1.00                0.0009664 2044
B/B            961 55.3 183 46.1 0.69  0.56  0.86
Overdominant
A/A-B/B       1076 61.9 227 57.2 1.00                0.0847411 2052
A/B            663 38.1 170 42.8 1.22  0.97  1.52
log-Additive
0,1,2         1739 81.4 397 18.6 0.72  0.61  0.85 0.0001259 2041
```

4.6   Post-GWAS visualization and interpretation

The main goal of genomic association studies is the identification of *any* variant that is significantly associated with phenotype differences. The analysis does not test a mechanistic hypothesis but discovers plausible biological associations that can guide into the mechanisms. The search for plausible mechanisms can be helped with the knowledge of whether a statistically significant SNP is within a protein-coding gene, intergenic, or close to a regulatory element (e.g., a methylation mark) in tissues or cell types that are relevant to the trait.

There are several databases that contain relevant biological information that can be used to biologically annotate SNPs. For instance, the GWAS Catalog contains information about SNPs associated with diseases at a genome-wide significant level, ENCODE databases describe chromatin states, epigenetic marks and transcription factor bindings of genomic sites, and GTEx data includes information about possible effects of SNPs on RNA expression. While there are a large number of data resources, we aim to illustrate how to use some that are commonly used. Section 11.4.2 in Chapter 11 also illustrates how to obtain further information from existing databases using *haploR* package.

We retake our asthma example illustrated in Section 4.2.3. While we did not find any significant association in the massive univariate analysis, we found haplotypic differences that were associated with asthma status in the region 34.5-35.0Mb of chromosome 7. We would then like to further inquire into the region and look for important biological links that help us interpret these results. In particular, we are interested in identifying SNPs that have been linked to any disease in a GWAS study, SNPs that regulate gene expression (eQTLs) or DNaseI hypersensitivity (dsQTL), available in GBrowse at `ht tp://eqtl.uchicago.edu/Home.html`. The information is available in the *brgedata* (downloaded from GBrowse on 9 Dec. 2017). Data from the GWAS catalog is accessible at `https://www.ebi.ac.uk/gwas/downloads`.

We illustrate how to plot a comprehensive annotation of the region. We first define a `GenomicRanges` *GenomicRanges*, then, using the *Gviz* package we annotate the eQTLs and dsQTL data tracks. Finally we use the function `gwcex2gviz`, from the Bioconductor package *gwascat*, to annotate GWAS results from the GWAS Catalog. All of these data are available from *brgedata* package. This is the code to reproduce Figure 4.15.

```
> library(gwascat)
> library(rtracklayer)
> library(Gviz)
>
> data(gwascatalog, package = "brgedata")
>
> gff <- "chr7_33000000_35000000.gff3"
```

```
>
> chr <- "chr7"
> ir <- IRanges(start=33e6, end=35e6)
> region <- GRanges(seqnames=chr, ir)
>
> pos <- snpInfo$chrom_start[snpInfo$chr_name=="7"]
> snps.loc <- GRanges(seqnames=chr,
+                     IRanges(start=pos,
+                     width=1))
>
> c7 <- import(gff)
>
> # annotate GWAS catalog
> annot <- gwcex2gviz(basegr = gwascatalog,
+                     contextGR=region, plot.it=FALSE)
>
> # annotate dsQTL
> c7ds <- c7[ which(mcols(c7)$type == "Degner_dsQTL") ]
> ds <- DataTrack( c7ds, chrom=chr, genome="hg19",
+                  data="score", name="dsQTL")
>
> # annotate eQTL
> c7e <- c7[ which(mcols(c7)$type == "Pickrell_eqtl") ]
> eq <- AnnotationTrack(c7e,  chrom=chr, genome="hg19",
+                       name="eQTL")
>
> # annotate SNPs from association study
> snps2 <- AnnotationTrack(snps.loc,  chrom=chr,
+                          genome="hg19", name="asthma SNPs")
>
> # Create plot
> displayPars(eq)$col <- "black"
> displayPars(ds)$col <- "red"
> displayPars(snps2)$col <- "darkgreen"
> integ <- list(annot[[1]], eq, ds, snps2, annot[[2]], annot[[3]])
> plotTracks(integ)
```

We observe (in Figure 4.15) that, although there is no dsQTL nor eQTL that map in the region (34.5-35.0 Mb), there are GWASs describing genome-wide significant associations with several traits. These traits can be extracted with

```
> traits <- subsetByOverlaps(gwascatalog,
+                            region)$DISEASE.TRAIT
> length(traits)
[1] 16
> head(traits, n=12)
 [1] "Cytomegalovirus antibody response"
 [2] "Sagittal craniosynostosis"
 [3] "Smooth-surface caries"
 [4] "Alzheimer's disease"
 [5] "IgG glycosylation"
 [6] "IgG glycosylation"
 [7] "Acute kidney injury in coronary artery bypass surgery (creatinine rise)"
 [8] "Post bronchodilator FEV1"
```

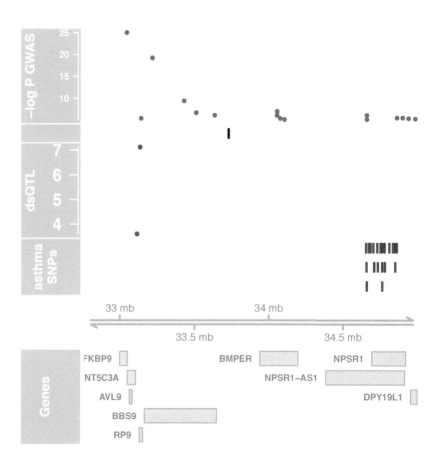

FIGURE 4.15
Genomic annotation of SNPs from asthma data example in chr7:34.5-35.0Mb
region joint with SNPs described in the GWAS catalog, eQTLs annotated in
Pickrell et al. (2010) and dsQTL from Degner et al. 2012.

```
 [9] "Post bronchodilator FEV1"
[10] "Post bronchodilator FEV1"
[11] "Post bronchodilator FEV1"
[12] "Post bronchodilator FEV1/FVC ratio"
```

Interestingly, we observe that there are some associations with intermediate phenotypes of asthma, such as forced expiration (FEV1), its ratio with forced vital capacity (FEV1/FVC) and IgG glycosylation. These observations offer additional biological evidence to the associations.

4.6.1 Genome-wide associations for imputed data

We use the package *CNVassoc*, explained in detail in the following chapter, to analyze SNP data when the SNPs have been imputed or genotyped with some degree of error

```
> library(CNVassoc)
```

A data example obtained from SNPTEST software (available on `http://www.stats.ox.ac.uk/~marchini/software/gwas/snptest.html`) has been incorporated in the *CNVassoc* package. IMPUTE [59] is a program to infer a set of unobserved SNPs from others that have been genotyped, using linkage disequilibrium and other information, usually from the HapMap project[1].

The data consists of 100 imputed SNPs on a set of 500 cases and 500 controls. For all of the SNPs, the probabilities of each genotype are given as provided by IMPUTE. The names of the SNPs, as well as the name of the disease, have been masked.

Let us start by loading the data. There are 2 data frames, one for cases and the other for controls

```
> data(SNPTEST, package="CNVassocData")
> dim(cases)
[1]  100 1505
> dim(controls)
[1]  100 1505
```

The structure of the data is as follows. Each row corresponds to a SNP. The first 3 columns are the SNP identification codes while the 4th and 5th are the alleles. The columns 6 through to the end provide the probabilities of each genotype, each group of 3 columns corresponds to one individual. This is, for instance, the information for cases

```
> cases[1:10,1:11]
  V1 V2   V3 V4 V5          V6            V7           V8          V9
1  1  1 1000  A  T 0.9959626125 0.0023620260 0.0016753615 0.992634932
2  2  2 2000  A  T 0.0765213302 0.0073893102 0.9160893596 0.027811741
```

---

[1]http://www.internationalgenome.org/category/hapmap/

```
3    3   3   3000   A   T  0.0050670931 0.0020722897 0.9928606172 0.009646064
4    4   4   4000   A   T  0.9920997158 0.0003108851 0.0075893991 0.012288000
5    5   5   5000   A   T  0.0048796013 0.0283927739 0.9667276249 0.990459821
6    6   6   6000   A   T  0.0029449045 0.9965970143 0.0004580812 0.993531065
7    7   7   7000   A   T  0.9844537961 0.0147126387 0.0008335652 0.003635098
8    8   8   8000   A   T  0.0002854996 0.0019421881 0.9977723123 0.005000345
9    9   9   9000   A   T  0.0052202003 0.0037747406 0.9910050592 0.003845385
10  10  10  10000   A   T  0.0145463505 0.9603995477 0.0250541018 0.010122825
                 V10            V11
1   0.0003516265 7.013442e-03
2   0.0086429180 9.635453e-01
3   0.0026860830 9.876679e-01
4   0.9815783730 6.133627e-03
5   0.0092745162 2.656632e-04
6   0.0023760942 4.092840e-03
7   0.9945822710 1.782631e-03
8   0.0024962428 9.925034e-01
9   0.0011333510 9.950213e-01
10  0.9898094554 6.771937e-05
```

and controls

```
> controls[1:10,1:8]
    V1 V2     V3 V4 V5           V6           V7           V8
1    1  1   1000  A  T  9.822425e-01 0.003358295 0.014399242
2    2  2   2000  A  T  1.333922e-02 0.969099360 0.017561421
3    3  3   3000  A  T  3.989599e-03 0.004256366 0.991754036
4    4  4   4000  A  T  3.406932e-03 0.007333515 0.989259553
5    5  5   5000  A  T  9.881081e-01 0.010474830 0.001417104
6    6  6   6000  A  T  3.595319e-03 0.990430376 0.005974305
7    7  7   7000  A  T  6.072451e-05 0.997494894 0.002444382
8    8  8   8000  A  T  6.322546e-03 0.006265613 0.987411841
9    9  9   9000  A  T  3.073608e-04 0.007901964 0.991790675
10  10 10  10000  A  T  9.748969e-03 0.978622828 0.011628203
```

For example, the first individual in the data set of cases has probabilities of 0.996, 0.0024 and 0.0017 of having the genotypes for the first SNP of AA, AT and TT respectively. And the second individual has a probabilities of 0.0278, 0.0086 and 0.9635 of having the genotypes for the second SNP of AA, AT and TT respectively. Cases and controls must have the same number of rows because the i-th row of cases and the i-th row of controls correspond to the same SNP.

Using *CNVassoc* requires some data preprocessing steps. The goal is to have one matrix of probabilities with 3 columns corresponding to the 3 genotypes and 1000 individuals (500 cases plus 500 controls), for each of the 100 SNPs. In our example, this can be done by simply:

```
> nSNP <- nrow(cases)
> probs <- lapply(1:nSNP, function(i) {
+    snpi.cases <- matrix(as.double(cases[i, 6:ncol(cases)]),
+                         ncol = 3, byrow = TRUE)
+ snpi.controls <- matrix(as.double(controls[i, 6:ncol(controls)]),
+
```

```
+                            ncol = 3, byrow = TRUE)
+ return(rbind(snpi.cases, snpi.controls))
+ })
```

Now, `probs` is a list of 100 components, each one containing the probability matrix of each SNP, and the first 500 rows of each matrix refer to the cases and the rest to the controls.

At this point, we use `multiCNVassoc` to perform an association test of each SNP with case-control status. But first, a case-control variable must be defined, which, in this example, will be a simple vector of 500 ones and 500 zeros.

```
> casecon <- rep(1:0, c(500, 500))
```

Now, we have the data ready to fit a model; for example, to compute the association P-value between every SNP and case-control status assuming an additive effect:

```
> pvals <- multiCNVassoc(probs, formula = "casecon ~ CNV",
+                         model = "add", num.copies = 0:2,
+                         cnv.tol = 0.001)
```

It is then necessary to correct for multiple tests:

```
> pvalsBH <- getPvalBH(pvals)
> head(pvalsBH)
  region       pval    pval.BH
1      1 0.29083371 0.8400958
2      3 0.13235295 0.8400958
3      5 0.08296301 0.8400958
4      6 0.18826664 0.8400958
5      7 0.24967318 0.8400958
6      9 0.30321197 0.8400958
```

A frequency tabulation of how many SNPs achieve different levels of significance is obtained by:

```
> table(cut(pvalsBH[, 2], c(-Inf, 1e-3, 1e-2, 0.05, 0.1, Inf)))

(-Inf,0.001] (0.001,0.01]  (0.01,0.05]   (0.05,0.1]   (0.1, Inf]
           0            0            2            7           91
```

From these results, no SNP appears to be associated with case-control status.

5

Genomic variant studies

CONTENTS

5.1   Chapter overview

In this chapter, we deal with genetic variants other than SNPs. As the proportion of genetic variation explained by SNPs for complex traits appears to be limited, more complex genetic variants are likely to play an important role in complex traits or disease susceptibility. We explain how SNP array data can be used to detect other structural variants (SV) such as copy number (CNV), genetic mosaicisms and polymorphic inversions. In particular, we discuss how CNVs, in Section (5.2), and mosaicisms in, Section (5.4), can be called using the log R ratio (LRR) and B-allele frequencies (BAF) of raw microarray data, and how polymorphic inversions can be called using SNP genotypes in Section (5.5). We demonstrate how to assess its association with quantitative and qualitative outcomes.

5.2 Copy number variants

In this section, we demonstrate how to detect regions with copy-number alterations, such as gains or losses of genetic material, using quantitative data obtained from aCGH or SNP arrays. We also illustrate how to incorporate the uncertainty derived from the calling of the copy-number status of individuals in association studies.

Microarrays measure the fluorescence light of tagged DNA segments that hybridize with the probes in the chip array. Determining the genotypes from two allele intensities depends on each technology and proprietary software. However, a common principle is to normalize the two allele signals for each SNP to obtain the SNP's Log R Ratio (LRR) and B Allele Frequency (BAF). The LRR is the normalized measure of total signal intensity and provides information relative to copy number or the number of alleles present in a biological sample at one SNP. Segmentation of LRR values along neighboring SNPs, that is the identification of regions with consistently high or low LRR across the SNPs, can be used to infer the DNA segments with probable copy number alterations. The BAF, on the other hand, is derived from the ratio of between intensities of the allele probes, measuring the allele proportion or allele composition of the sample. In normal samples, the expected values are 0 and 1 for homozygous genotypes (AA, BB) and 0.5 for heterozygous genotypes (AB) (Figure 5.2, normal event). Biological samples with different genetic content among cells will show a split in the BAF of heterozygous SNPs that will extend in chromosomal mosaicisms with no apparent LRR differences (Figure 5.2),[126].

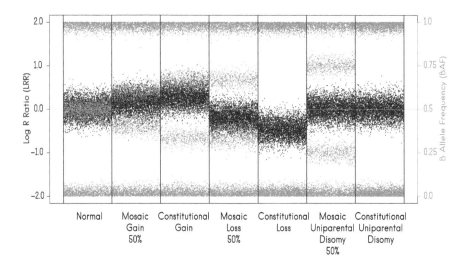

FIGURE 5.1

Different chromosomal configurations in a genotyping array. Each interrogated SNP is represented by gray and black dots. Black dots represent LRR, and gray dots BAF. The different configurations start (left panel) with a hypothetical normal diploid sample followed by a different constitutional CNV gain, loss and uniparental disomy. The figure also illustrates a mosaic event representation with a 50% cellularity corresponding to each type of CNV.

5.2.1 CNV calling

There are several packages to perform CNV calling using SNP array data. These include *CBS* [105], *R-GADA* [111], *crlmm* [134], *copynumber* [104] among others. Various databases provide genomic data where CNV calling has been performed, see for instance the package *RTCGA.cnv*. We illustrate how to call CNV with the *R-GADA* package.

Segmentation of LRR values along the chromosome together with BAF information are used to identify the regions with probable copy number alterations. Using different tools, the required information can be obtained from raw microarray data. Affymetrix data (.CEL files) can be processed using the Birdseed v2 algorithm (http://software.broadinstitute.org/cancer /software/genepattern/affymetrix-snp6-copy-number-inference- pipeline), Affymetrix power tools (https://www.affymetrix.com/suppo rt/developer/powertools/changelog/index.html) or *affy2sv* R package [52]. Illumina data (.idat files) can be processed with Genome Studio software (https://www.illumina.com/techniques/microarrays/array-da ta-analysis-experimental-design/genomestudio.html). The *crlmm* Bioconductor package can also extract LRR and BAF [134].

These tools can be used to configure data in PennCNV format (http: //penncnv.openbioinformatics.org/en/latest/user-guide/input/), which is required for the different methods that we discuss. In the PennCNV format, there is a text file for each subject with genome-wide information about SNP, chromosome, position, LRR, BAF, and genotype, as follows

```
Name        Chr    Position    Log.R.Ratio B.Allele.Freq    GType
rs758676     7     12878632        0.1401        0.4977        AB
rs3916934   13    103143536        0.3934        0.4610        AA
rs2711935    4     38838852       -0.1091        0.0026        AA
rs17126880   1     64922104        0.0478        0.9910        AA
rs12831433  12     4995220        -0.1661        0.0000        AA
...
```

The *R-GADA* package uses this format to perform CNV calling. It can be installed by:

```
> devtools::install_github("isglobal-brge/R-GADA")
```

The package requires a strict directory structure of the subjects' data in the working directory for performing the analysis. The user needs to set up a directory `rawData` with the PennCNV files. This is an example of how data must be organized before analysis

```
|-- rawData
|    |-- CASE369.txt
|    |-- CASE371.txt
|    |-- CASE377.txt
```

```
|   |-- CONTROL152.txt
|   |-- CONTROL191.txt
```

It is important that all files included in the folder `rawData` belong to the same type of array because annotation data is obtained from one of the files and shared across all others. Samples that have been analyzed with different platforms should be processed in other paths that contain their own `rawData` directories.

Let us illustrate how to perform CNV calling using the data available in the `brgedata` package. We start by using the loaded `madData` to create the required files for analysis

```
> ss1 <- system.file("extdata/madData", package="brgedata")
> dir.create("rawData")
> ss2 <- "rawData"
> files <- list.files(ss1)
> file.copy(file.path(ss1,files), ss2)
[1] FALSE FALSE FALSE FALSE FALSE
```

We can check that the files have been copied correctly

```
> dir("rawData")
[1] "CASE369.txt"     "CASE371.txt"     "CASE377.txt"     "CONTROL152.txt"
[5] "CONTROL191.txt"
```

We then import raw data into R using `setupParGADA` function. This function assumes by default that the first three columns contain information about SNP name, chromosome and position; although these can be changed (see `?setupParGADA`).

```
> library(gada)


> cnv.call <- setupParGADA(log2ratioCol = 4,
+                          BAFcol = 5)
Creating object with annotation data ...done

Creating objects of class setupGADA for all input files...
  Applying setupGADAIllumina for 5 samples ...
  Importing array:  CASE369.txt ...    Array # 1 ...done
  Importing array:  CASE371.txt ...    Array # 2 ...done
  Importing array:  CASE377.txt ...    Array # 3 ...done
  Importing array:  CONTROL152.txt ...    Array # 4 ...done
  Importing array:  CONTROL191.txt ...    Array # 5 ...done
  Applying setupGADAIllumina for 5 samples ... done
Creating objects of class setupGADA for all input files... done
> cnv.call
[1] "C:/Juan/CREAL/GitHub/multiomic_book/data"
attr(,"class")
[1] "parGADA"
attr(,"labels.samples")
[1] "CASE369"    "CASE371"    "CASE377"    "CONTROL152" "CONTROL191"
attr(,"Samples")
[1] 5
```

setupParGADA creates an object of *parGADA* class. The LRR and BAF columns are specified through the arguments BAFcol and log2ratioCol, respectively.

Segmentation of the individuals' genomic data is performed by two consecutive algorithms implemented in two separate R functions. The first function (parSBL) uses sparse Bayesian learning (SBL) to discover the most likely positions and magnitudes for a segment, i.e. the breakpoints. The SBL model is governed by a hierarchical Bayesian prior, which is uninformative with respect to the location and magnitude of the copy number changes but restricts the total number of breakpoints. Sensitivity, given by the maximum breakpoint sparseness, is controlled by the hyperparameter aAlpha.

The second function (parBE) is an algorithm that uses a backward elimination (BE) strategy to rank the statistical significance of each breakpoint obtained from SBL (parameter T). It sequentially removes the least significant breakpoints estimated by the SBL model and allows a flexible adjustment of the False Discovery Rate (FDR). Table 5.1 provides the values of both aAlpha and T arguments that must be considered to increase sensitivity and specificity.

(higher sensitivity , higher FDR)	$\longleftrightarrow$	$(a_\alpha = 0.2, T > 3)$
	$\longleftrightarrow$	$(a_\alpha = 0.5, T > 4)$
(lower sensitivity , lower FDR)	$\longleftrightarrow$	$(a_\alpha = 0.8, T > 5)$

TABLE 5.1
Parameters used in GADA to control sensitivity and false discovery rate (FDR).

In our example, we call SBL with the $\alpha = 0.8$

```
> parSBL(cnv.call, estim.sigma2=TRUE, aAlpha=0.8)
Creating SBL directory ...done
Retrieving annotation data ...done
Segmentation procedure for 5 samples ...
   Array # 1 ...     The estimated sigma2 = 0.01931369
   Array # 1 ...done
   Array # 2 ...     The estimated sigma2 = 0.02835627
   Array # 2 ...done
   Array # 3 ...     The estimated sigma2 = 0.02604537
   Array # 3 ...done
   Array # 4 ...     The estimated sigma2 = 0.03035296
   Array # 4 ...done
   Array # 5 ...     The estimated sigma2 = 0.04091223
   Array # 5 ...done
Segmentation procedure for 5 samples ...done
```

the function updates the cnv.call with the segmentation results which are then passed to the backward elimination algorithm. In our example we use $T = 6$ and minimum length of 25 in the parBE function.

```
> parBE(cnv.call, T=6, MinSegLen=25)
Retrieving annotation data ...done
Backward elimination procedure for 5 samples ...
   Array # 1 ... ----------------------------------------
Sparse Bayesian Learnig (SBL) algorithm
Backward Elimination procedure with T=6 and minimun length size=25
 Number of segments =  108
 Base Amplitude of copy number 2: chr 1:22:0.0151, X=-0.058, Y=0.0921
   Array # 2 ... ----------------------------------------
Sparse Bayesian Learnig (SBL) algorithm
Backward Elimination procedure with T=6 and minimun length size=25
 Number of segments =  71
 Base Amplitude of copy number 2: chr 1:22:0.0171, X=-0.0592, Y=0.0529
   Array # 3 ... ----------------------------------------
Sparse Bayesian Learnig (SBL) algorithm
Backward Elimination procedure with T=6 and minimun length size=25
 Number of segments =  106
 Base Amplitude of copy number 2: chr 1:22:-0.0311, X=-0.1395, Y=-0.0916
   Array # 4 ... ----------------------------------------
Sparse Bayesian Learnig (SBL) algorithm
Backward Elimination procedure with T=6 and minimun length size=25
 Number of segments =  163
 Base Amplitude of copy number 2: chr 1:22:-0.0415, X=-0.1804, Y=-0.5418
   Array # 5 ... ----------------------------------------
Sparse Bayesian Learnig (SBL) algorithm
Backward Elimination procedure with T=6 and minimun length size=25
 Number of segments =  70
 Base Amplitude of copy number 2: chr 1:22:-0.0132, X=-0.0941, Y=0.07
Backward elimination procedure for 5 samples ...done
```

The results are stored in a folder called **SBL** that are accessed through the *parGADA* object `cnv.call`, they can be summarized with the generic function **summary**. The results specify the length of the CNVs detected and their limits

```
> summ.cnvs <- summary(cnv.call)
> summ.cnvs


---------------------------------------------
Summary results for 5 individuals
---------------------------------------------
NOTE: 32 segments with length not in the range 0-Inf bases
 and with mean log2ratio in the range (-0.02,0.01) have been discarded

Number of Total Segments:
 # segments Gains     % Losses     %
        87     26 29.9     61 70.1

Summary of length of segments:
     Min.   1st Qu.    Median      Mean   3rd Qu.      Max.
    69746  22640209  62905147  65240559  96093594 185546720

Number of Total Segments by chromosome:
             segments Gains Losses
Chromosome 1        7     3      4
Chromosome 2        2     1      1
```

```
Chromosome 3      2    1    1
Chromosome 4      7    0    7
Chromosome 5      4    1    3
Chromosome 6     14    1   13
Chromosome 7      5    3    2
Chromosome 8      6    3    3
Chromosome 9      3    0    3
Chromosome 10     4    3    1
Chromosome 11     7    1    6
Chromosome 12     3    2    1
Chromosome 13     2    0    2
Chromosome 14     5    2    3
Chromosome 15     4    2    2
Chromosome 16     3    2    1
Chromosome 17     0    0    0
Chromosome 18     1    0    1
Chromosome 19     2    1    1
Chromosome 20     3    0    3
Chromosome 21     3    0    3
Chromosome 22     0    0    0
```

We can filter the length of the CNVs with the parameter `length.base`. Results are filtered by the mean LRR within given limits. The LRR limits in our example are given by

```
> findNormalLimits(cnv.call)
[1] -0.02365792  0.01383900
```

That is a segment corresponding to 2 copy gains. By default, these limits are estimated using a threshold approach to classify segments into gain and loss state. The threshold is automatically estimated using the X chromosome of a normal population that includes males (XY) and females (XX) but can be modified with the argument `threshold`. This example changes the length of the segment to avoid large CNVs that could be false results (in bases from 500 up to 10^6). The limits of normal segments could also be changed with the parameter `threshold`.

```
> summ2.cnvs <- summary(cnv.call, length.base=c(500,1e6))
> summ2.cnvs

---------------------------------------------
Summary results for 5 individuals
---------------------------------------------
NOTE: 142 segments with length not in the range 500-1e+06 bases
 and with mean log2ratio in the range (-0.02,0.01) have been discarded

Number of Total Segments:
 # segments Gains  % Losses  %
          2    1 50      1 50

Summary of length of segments:
   Min. 1st Qu.  Median    Mean 3rd Qu.    Max.
  69746  234720  399694  399694  564667  729641
```

```
Number of Total Segments by chromosome:
              segments Gains Losses
Chromosome 1         0     0      0
Chromosome 2         0     0      0
Chromosome 3         0     0      0
Chromosome 4         0     0      0
Chromosome 5         0     0      0
Chromosome 6         0     0      0
Chromosome 7         0     0      0
Chromosome 8         0     0      0
Chromosome 9         0     0      0
Chromosome 10        1     1      0
Chromosome 11        0     0      0
Chromosome 12        0     0      0
Chromosome 13        0     0      0
Chromosome 14        0     0      0
Chromosome 15        0     0      0
Chromosome 16        0     0      0
Chromosome 17        0     0      0
Chromosome 18        1     0      1
Chromosome 19        0     0      0
Chromosome 20        0     0      0
Chromosome 21        0     0      0
Chromosome 22        0     0      0
```

We can obtain a *GenomicRanges* with the detected alterations with gains (1) and losses (-1) as indicated in the column `State`. The column `LenProbe` encodes the number of probes (SNPs) that are in the segment (e.g. range) while the `MeanAmp` column encodes the mean amplitude of the segment, expected to be different from 0 for significant copy number alterations.

```
> library(GenomicRanges)
> cnvs <- getCNVs(summ.cnvs)
> cnvs
GRanges object with 112 ranges and 4 metadata columns:
          seqnames              ranges strand |  LenProbe   MeanAmp
             <Rle>           <IRanges>  <Rle> | <integer> <numeric>
    [1]       chr1  12878632-30710941      * |      1267      0.04
    [2]       chr1  6419322-125421011      * |      1874      0.03
    [3]       chr1 124791023-162916253     * |      5018      0.02
    [4]       chr1  36662686-86461049      * |        86     -0.28
    [5]       chr1  12878632-85912483      * |      1268     -0.03
    ...       ...                 ...    ... .       ...       ...
  [108]       chrY 68213861-107449195      * |        42     -0.71
  [109]       chrY 23076329-117822956      * |       110     -0.72
  [110]       chrY 53866640-112459430      * |        26     -0.29
  [111]       chrY  4353400-10939173       * |        33     -0.87
  [112]       chrY 131128810-159666866     * |        44     -0.18
              State      sample
          <numeric> <character>
    [1]           1     CASE369
    [2]           1     CASE369
    [3]           1     CASE369
    [4]          -1     CASE371
    [5]          -1     CASE377
```

```
     ...          ...          ...
   [108]          -1   CONTROL152
   [109]          -1   CONTROL152
   [110]           1   CONTROL152
   [111]          -1   CONTROL152
   [112]          -1   CONTROL191
   -------
   seqinfo: 22 sequences from an unspecified genome; no seqlengths
```

Alterations for a given sample (CASE377) can be retrieved

```
> subset(cnvs, sample=="CASE377")
GRanges object with 17 ranges and 4 metadata columns:
          seqnames             ranges strand |  LenProbe   MeanAmp      State
             <Rle>          <IRanges>  <Rle> | <integer> <numeric>  <numeric>
     [1]      chr1  12878632-85912483      * |      1268     -0.03         -1
     [2]      chr1 18678151-161671331      * |      3806     -0.03         -1
     [3]      chr4 43356005-138883986      * |       198     -0.18         -1
     [4]      chr6  2938029-16221861      * |       549     -0.07         -1
     [5]      chr6 12000608-161505196      * |        29     -0.41         -1
     ...       ...                ...    ... .       ...       ...        ...
    [13]     chr20 73286914-110049028      * |       738     -0.04         -1
    [14]     chr21 115763225-241564328     * |      3221     -0.03         -1
    [15]      chrX 101394185-136363752     * |        27     -0.56         -1
    [16]      chrX  34610708-80037722      * |        46     -0.46         -1
    [17]      chrY 131128810-159666866     * |        44     -0.38         -1
              sample
           <character>
     [1]      CASE377
     [2]      CASE377
     [3]      CASE377
     [4]      CASE377
     [5]      CASE377
     ...          ...
    [13]      CASE377
    [14]      CASE377
    [15]      CASE377
    [16]      CASE377
    [17]      CASE377
   -------
   seqinfo: 22 sequences from an unspecified genome; no seqlengths
```

as alterations in a specific chromosome (6)

```
> subset(cnvs, seqnames=="chr6")
GRanges object with 14 ranges and 4 metadata columns:
          seqnames             ranges strand |  LenProbe   MeanAmp      State
             <Rle>          <IRanges>  <Rle> | <integer> <numeric>  <numeric>
     [1]      chr6  3018937-65924084      * |       170     -0.32         -1
     [2]      chr6 12000608-161505196      * |        29      0.27          1
     [3]      chr6  3018937-96405016      * |        73     -0.35         -1
     [4]      chr6  44534480-62137546      * |       140     -0.34         -1
     [5]      chr6 68611259-165270466      * |       129     -0.17         -1
     ...       ...                ...    ... .       ...       ...        ...
    [10]      chr6  67279023-69974616      * |        31     -0.38         -1
    [11]      chr6 30730193-122782661      * |      9051     -0.03         -1
```

```
[12]    chr6 67279023-205395955    * |    37    -0.35    -1
[13]    chr6   5116355-10327073    * |    56    -0.26    -1
[14]    chr6 33653283-165270466    * |   335    -0.18    -1
             sample
          <character>
  [1]     CASE369
  [2]     CASE369
  [3]     CASE371
  [4]     CASE371
  [5]     CASE371
  ...         ...
 [10]     CASE377
 [11]  CONTROL152
 [12]  CONTROL152
 [13]  CONTROL191
 [14]  CONTROL191
 -------
 seqinfo: 22 sequences from an unspecified genome; no seqlengths
```

We can visualize the results using Bioconductor's *Gviz* package [48]. The function `plotCNVs` in *R-GADA* is used to create Figure 5.2. The figure shows the CNVs (gains and losses) detected in the region chr8:45-60Mb for the 5 samples previously analyzed. Two of them have gains while another one has a loss in that region. Gene symbols are annotated by using Bioconductor's *Homo.sapiens* package and setting the argument `drawGenes` equal to TRUE:

```
> library(Gviz)
> library(Homo.sapiens)


> rr <- GRanges("chr8:45.0e6-60.0e6")
> plotCNVs(cnvs, range=rr, drawGenes = TRUE,
+          col.cnvs = c("darkgray", "lightgray"))
```

We now illustrate how to perform an association analysis between each CNV (range) and a factor variable (e. g. case/control). We thus compare the number of cases with the number of controls that have an alteration in a given region. Therefore, we need to separate cases and controls. We first add a factor variable corresponding to health status to the `cnvs` results

```
> casecont <- rep("control", length(cnvs))
> casecont[grep("CASE", cnvs$sample)] <- "case"
> cnvs$casecont <- as.factor(casecont)
```

and then select the alterations for cases and controls

```
> cnvs.cases <- subset(cnvs, casecont=="case")
> cnvs.controls <- subset(cnvs, casecont=="control")
```

For testing whether there is any CNV (e.g. range or segment) whose differences in the number of cases and controls are significantly different, we first compute the number of cases and controls in a given range

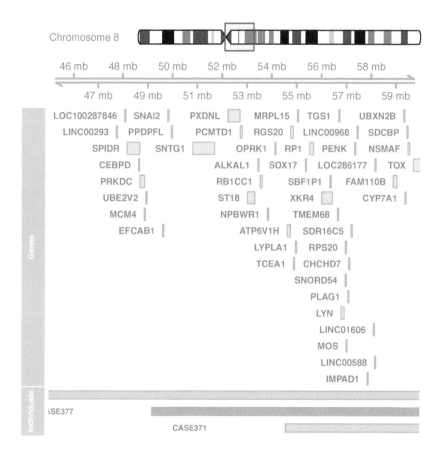

FIGURE 5.2
CNV gains (dark gray) and losses (light gray) in the region chr8:45-60Mb of
samples of *brgedata* package.

```
> ss <- unique(cnvs.controls$sample)
> cnvs.controls.list <- GRangesList(lapply(ss, function(x)
+   subset(cnvs.controls, sample==x)))
> count.controls <- countOverlaps(cnvs, cnvs.controls.list)
>
> ss <- unique(cnvs.cases$sample)
> cnvs.cases.list <- GRangesList(lapply(ss, function(x)
+   subset(cnvs.cases, sample==x)))
> count.cases <- countOverlaps(cnvs, cnvs.cases.list)
> cnvs$counts <- cbind(count.controls, count.cases)
> cnvs[, c("counts")]
GRanges object with 112 ranges and 1 metadata column:
          seqnames               ranges strand |    counts
             <Rle>            <IRanges>  <Rle> |  <matrix>
     [1]      chr1   12878632-30710941      * |       1:2
     [2]      chr1    6419322-125421011     * |       1:3
     [3]      chr1  124791023-162916253     * |       0:2
     [4]      chr1   36662686-86461049      * |       1:3
     [5]      chr1   12878632-85912483      * |       1:3
     ...       ...                  ...    ... .       ...
   [108]      chrY   68213861-107449195     * |       1:1
   [109]      chrY   23076329-117822956     * |       1:2
   [110]      chrY   53866640-112459430     * |       1:1
   [111]      chrY    4353400-10939173      * |       1:0
   [112]      chrY  131128810-159666866     * |       2:3
   -------
   seqinfo: 22 sequences from an unspecified genome; no seqlengths
```

We can see that one case and zero controls have an alteration in the first segment. For the second segment, for instance, overlapping segments were found for one case and one control

```
> subsetByOverlaps(cnvs, cnvs[2,])
GRanges object with 7 ranges and 6 metadata columns:
          seqnames               ranges strand |  LenProbe   MeanAmp     State
             <Rle>            <IRanges>  <Rle> | <integer> <numeric> <numeric>
     [1]      chr1   12878632-30710941      * |      1267      0.04         1
     [2]      chr1    6419322-125421011     * |      1874      0.03         1
     [3]      chr1  124791023-162916253     * |      5018      0.02         1
     [4]      chr1   36662686-86461049      * |        86     -0.28        -1
     [5]      chr1   12878632-85912483      * |      1268     -0.03        -1
     [6]      chr1   18678151-161671331     * |      3806     -0.03        -1
     [7]      chr1   18678151-111916489     * |      5020     -0.04        -1
                 sample casecont   counts
            <character> <factor> <matrix>
     [1]      CASE369       case      1:2
     [2]      CASE369       case      1:3
     [3]      CASE369       case      0:2
     [4]      CASE371       case      1:3
     [5]      CASE377       case      1:3
     [6]      CASE377       case      1:3
     [7]   CONTROL152    control      1:3
   -------
   seqinfo: 22 sequences from an unspecified genome; no seqlengths
```

We use a binomial or Fisher's test to compare the differences in the proportion of cases and controls with an alteration in a given segment. Retrieving the total number of individuals

```
> agg <- aggregate(rep(1, length(cnvs)),
+                  by=list(casecont=cnvs$casecont,
+                                sample=cnvs$sample),
+                  FUN=sum)
> agg
  casecont     sample  x
1     case    CASE369 20
2     case    CASE371 17
3     case    CASE377 17
4  control CONTROL152 43
5  control CONTROL191 15
> n <- table(agg$casecont)
> n

   case control
      3       2
```

The genome-wide association analysis of CNVs follows from the massive application of Fisher's test on the cnv results

```
> testCNV <- function(x, n) {
+     tt <- matrix(c(x[1], n[1] - x[1], x[2], n[2] - x[2]), ncol=2)
+     ans <- try(fisher.test(tt), TRUE)
+     if (inherits(ans, "try-error"))
+         out <- NA
+     else
+         out <- ans$p.value
+     out
+ }
>
> cnvs$pvalue <- apply(cnvs$counts, 1, testCNV, n=n)
> cnvs$BH <- p.adjust(cnvs$pvalue, method="BH")
> cnvs[,c("counts", "pvalue", "BH")]
GRanges object with 112 ranges and 3 metadata columns:
          seqnames                 ranges strand |   counts    pvalue        BH
             <Rle>              <IRanges>  <Rle> | <matrix> <numeric> <numeric>
    [1]       chr1   12878632-30710941       * |      1:2       0.4         1
    [2]       chr1    6419322-125421011      * |      1:3      <NA>      <NA>
    [3]       chr1 124791023-162916253       * |      0:2       0.1         1
    [4]       chr1   36662686-86461049       * |      1:3      <NA>      <NA>
    [5]       chr1   12878632-85912483       * |      1:3      <NA>      <NA>
    ...        ...                   ...    ... .      ...       ...       ...
  [108]       chrY  68213861-107449195      * |      1:1         1         1
  [109]       chrY  23076329-117822956      * |      1:2       0.4         1
  [110]       chrY  53866640-112459430      * |      1:1         1         1
  [111]       chrY    4353400-10939173      * |      1:0         1         1
  [112]       chrY 131128810-159666866      * |      2:3      <NA>      <NA>
  -------
  seqinfo: 22 sequences from an unspecified genome; no seqlengths
```

As it can be seen, we adjust for multiple testing using the Benjamini–Hochberg method in the p.adjust function. Note that Fisher's test is suitable for a small number of samples while the binomial test is recommended for moderate/large sample sizes, which can be applied with the function `prop.test`.

We now illustrate genome-wide association analysis of CNVs for publicly available data of CNV calling. We use data from the TCGA project, available from the *RTCGA.CNV* package. The data is given in a similar format of that provided by *R-GADA*. Data for breast cancer (BRCA) is loaded as follows

```
> library(RTCGA.CNV)
```

```
> head(BRCA.CNV)
                        Sample Chromosome     Start       End Num_Probes
1 TCGA-3C-AAAU-10A-01D-A41E-01          1   3218610  95674710      53225
2 TCGA-3C-AAAU-10A-01D-A41E-01          1  95676511  95676518          2
3 TCGA-3C-AAAU-10A-01D-A41E-01          1  95680124 167057183      24886
4 TCGA-3C-AAAU-10A-01D-A41E-01          1 167057495 167059336          3
5 TCGA-3C-AAAU-10A-01D-A41E-01          1 167059760 181602002       9213
6 TCGA-3C-AAAU-10A-01D-A41E-01          1 181603120 181609567          6
  Segment_Mean
1       0.0055
2      -1.6636
3       0.0053
4      -1.0999
5      -0.0008
6      -1.2009
```

We select segments with copy number losses such that LRR mean value, given by **Segment_Mean**, is lower than $\log_2(1/2)$. Gains are selected such that **Segment_Mean** $> \log_2(3/2)$. We also select segments with number of probes between 5 and 500:

```
> dim(BRCA.CNV)
[1] 284458      6
> BRCA.CNV <- subset(BRCA.CNV, (Segment_Mean > log2(3/2) |
+                    Segment_Mean < log2(1/2)) &
+                    Num_Probes >= 5 & Num_Probes < 500)
> dim(BRCA.CNV)
[1] 27881      6
```

We then create a *GenomicRanges* object for this data

```
> state <- ifelse(BRCA.CNV$Segment_Mean>0, 1, -1)
>
> BRCA.CNV$Chromosome <- paste0("chr", BRCA.CNV$Chromosome)
> BRCA.CNV$Chromosome[BRCA.CNV$Chromosome=="chr23"] <- "chrX"
> brca.gr <- GRanges(seqnames = BRCA.CNV$Chromosome,
+                    ranges = IRanges(start = BRCA.CNV$Start,
+                                       end = BRCA.CNV$End),
+                    LenProbe = BRCA.CNV$Num_Probes,
+                    MeanAmp = BRCA.CNV$Segment_Mean,
+                    State = state,
+                    sample = BRCA.CNV$Sample)
>
```

```
> brca.gr$tumor <- rep(NA, length(brca.gr))
> brca.gr$tumor[grep("10A", brca.gr$sample)] <- "normal"
> brca.gr$tumor[grep("01A", brca.gr$sample)] <- "tumor"
> brca.gr <- brca.gr[!is.na(brca.gr$tumor), ]
>
> brca.gr$sample <- substr(brca.gr$sample, 1, 12)
> brca.gr
GRanges object with 26951 ranges and 5 metadata columns:
                seqnames              ranges strand | LenProbe   MeanAmp
                   <Rle>           <IRanges>  <Rle> | <integer> <numeric>
       [1]         chr1 181603120-181609567      * |        6   -1.2009
       [2]         chr2   51517041-51524666      * |       11   -1.1753
       [3]         chr3   89268644-89268812      * |        6    0.8839
       [4]        chr10   33871088-33873361      * |        5   -1.2802
       [5]        chr14   24908731-25015355      * |       10    0.6698
       ...          ...                 ...    ... .      ...       ...
   [26947]        chr11   73552530-73878688      * |      174    1.9592
   [26948]        chr11   74853813-74901601      * |       40    2.1277
   [26949]        chr11 123001084-123513213      * |      403    1.8318
   [26950]        chr11 123530470-123541653      * |       14    2.0101
   [26951]        chr16     653459-2025678      * |      356    0.6939
                   State       sample       tumor
               <numeric>  <character> <character>
       [1]          -1 TCGA-3C-AAAU      normal
       [2]          -1 TCGA-3C-AAAU      normal
       [3]           1 TCGA-3C-AAAU      normal
       [4]          -1 TCGA-3C-AAAU      normal
       [5]           1 TCGA-3C-AAAU      normal
       ...         ...          ...         ...
   [26947]           1 TCGA-Z7-A8R6       tumor
   [26948]           1 TCGA-Z7-A8R6       tumor
   [26949]           1 TCGA-Z7-A8R6       tumor
   [26950]           1 TCGA-Z7-A8R6       tumor
   [26951]           1 TCGA-Z7-A8R6       tumor
   -------
   seqinfo: 23 sequences from an unspecified genome; no seqlengths
```

For visualization with *Gviz* chromosome names must be annotated in the UCSC format, starting with the "chr" string. The labels 10A and 01A in the TCGA IDs correspond to either a normal or cancerous sample, respectively. The ID of an individual is the first 12 characters of the individual's TCGA ID.

The total number of samples is

```
> length(unique(brca.gr$sample))
[1] 1047
```

Figure 5.3 shows the CNVs reported in the genomic region chr7:5.0-8.0Mb

```
> rr <- GRanges("chr7:5e6-8e6")
> plotCNVs(brca.gr, range = rr, drawGenes = TRUE,
+          col.cnvs=c("darkgray", "lightgray"))
```

To perform genome-wide association analysis of CNVs, we count the CNVs

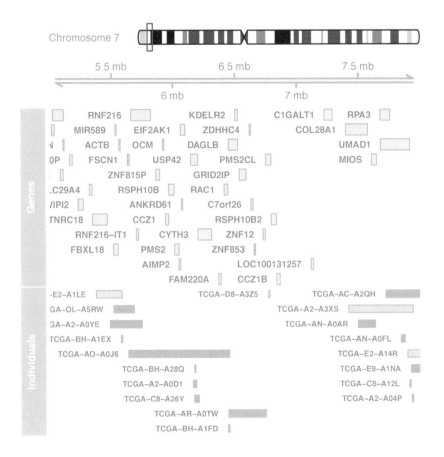

FIGURE 5.3
CNVs gain (dark gray) and losses (light gray) in the region chr7:5.0-8.0Mb of
samples belonging to the BRCA dataset from TCGA.

of cases and controls. The function `getCounts` of *R-GADA* package does the job. The argument `group` is used to indicate the grouping variable. Finally, the association analysis is performed with `testCNV` and adjustment for multiple comparisons with `p.adjust`

```
> ans <- getCounts(brca.gr, "tumor")
> brca.gr$counts <- ans$counts
> n <- ans$n
>
> brca.gr$pvalue <- apply(brca.gr$counts, 1, testCNV, n=n)
> brca.gr$BH <- p.adjust(brca.gr$pvalue, method="BH")
> brca.gr.sig <- subset(brca.gr[,c("counts", "pvalue", "BH")],
+                       brca.gr$BH < 0.01)
> brca.gr.sig
GRanges object with 7362 ranges and 3 metadata columns:
          seqnames               ranges strand |   counts
             <Rle>            <IRanges>  <Rle> | <matrix>
     [1]      chr1 149898951-150333087      * |     1:30
     [2]      chr1 150335347-151170789      * |     0:30
     [3]      chr1 154984468-155005476      * |     0:21
     [4]      chr1 155006954-155050185      * |     0:22
     [5]     chr11   60943071-61361689      * |     0:16
     ...       ...                  ...    ... .      ...
  [7358]     chr11   69559601-69650887      * |     0:66
  [7359]     chr11   70695866-71002410      * |     1:53
  [7360]     chr11   73552530-73878688      * |     0:22
  [7361]     chr11   74853813-74901601      * |     0:21
  [7362]     chr16     653459-2025678      * |     1:16
                          pvalue                      BH
                       <numeric>               <numeric>
     [1] 9.17196804835805e-07 6.85696840142296e-06
     [2] 4.02900812259604e-08 3.69842635940347e-07
     [3] 6.60494300658908e-06 4.05859140379805e-05
     [4] 3.64039791063025e-06 2.34223634083979e-05
     [5] 0.000138664831551331  0.000654493148010495
     ...                   ...                     ...
  [7358] 2.94813576696704e-17 9.70149048297054e-16
  [7359] 1.24305369365935e-12 2.43117126979776e-11
  [7360] 3.64039791063025e-06 2.34223634083979e-05
  [7361] 6.60494300658908e-06 4.05859140379805e-05
  [7362]  0.00101824759029954   0.00399167866271459
  -------
  seqinfo: 23 sequences from an unspecified genome; no seqlengths
```

In this example, we observe that in the region chr16:0.65-2.02Mb there are 1 normal and 16 tumor samples with CNV that leads to a corrected *P*-value $= 3.99 \times 10^{-3}$), see Figure 5.4 to get information about the genes within this region, which is obtained by

```
> rr <- GRanges("chr16:653459-2025678")
> subsetByOverlaps(brca.gr, rr)
GRanges object with 23 ranges and 8 metadata columns:
       seqnames             ranges strand | LenProbe   MeanAmp       State
          <Rle>          <IRanges>  <Rle> | <integer> <numeric>   <numeric>
```

```
  [1]    chr16 2010857-2011126      * |         5   -1.0992          -1
  [2]    chr16  988889-1161810      * |        45    0.9665           1
  [3]    chr16 1179031-1665665      * |        95    1.3482           1
  [4]    chr16 1665723-2568806      * |       309    1.0214           1
  [5]    chr16 1886249-2270113      * |       111    0.6972           1
  ...      ...            ...     ... .       ...       ...         ...
 [19]    chr16 1161810-2963528      * |       493    0.6538           1
 [20]    chr16   653459-655262      * |         6    0.7119           1
 [21]    chr16  653459-1007763      * |        69    0.9884           1
 [22]    chr16   653459-668980      * |         8    0.8168           1
 [23]    chr16  653459-2025678      * |       356    0.6939           1
                sample       tumor    counts              pvalue
           <character> <character> <matrix>           <numeric>
  [1] TCGA-A2-A0ES      normal      1:10  0.0287901477460552
  [2] TCGA-A7-A425      tumor       0:10  0.00655269845405028
  [3] TCGA-A7-A425      tumor        0:9  0.00646126064199843
  [4] TCGA-A7-A425      tumor       1:11  0.0164839199100839
  [5] TCGA-A8-A070      tumor       1:11  0.0164839199100839
  ...          ...        ...        ...                 ...
 [19] TCGA-D8-A142      tumor       1:12  0.00942675451318419
 [20] TCGA-D8-A1X9      tumor       0:10  0.00655269845405028
 [21] TCGA-E9-A1R4      tumor       0:12  0.00176896242355705
 [22] TCGA-E9-A1RC      tumor       0:10  0.00655269845405028
 [23] TCGA-Z7-A8R6      tumor       1:16  0.00101824759029954
                     BH
              <numeric>
  [1] 0.0694587120136008
  [2] 0.0194089214237948
  [3] 0.0194089214237948
  [4]  0.043994664834291
  [5]  0.043994664834291
  ...                ...
 [19] 0.0276844786841917
 [20] 0.0194089214237948
 [21] 0.00649623673504729
 [22] 0.0194089214237948
 [23] 0.00399167866271459
 -------
 seqinfo: 23 sequences from an unspecified genome; no seqlengths
```

```
> rr <- GRanges("chr16:653459-2025678")
> plotCNVs(brca.gr, range = rr, group="tumor", drawGenes = TRUE,
+          col.group = c("darkgray", "lightgray"))
```

The analyses can be performed for gains or losses independently, subsetting the object brca.gr with the State column

```
> brca.gr.gains <- brca.gr[brca.gr$State==1,]
> brca.gr.loses <- brca.gr[brca.gr$State==-1,]
> length(brca.gr)
[1] 26951
> length(brca.gr.gains)
[1] 22202
> length(brca.gr.loses)
[1] 4749
```

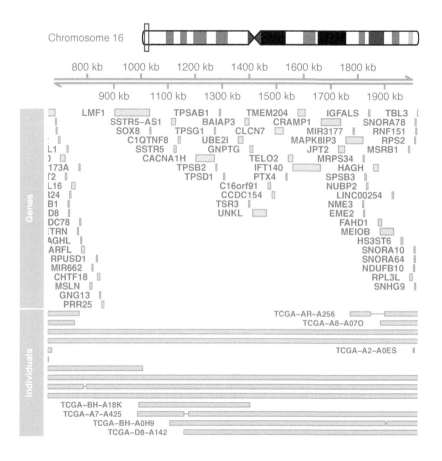

FIGURE 5.4
CNVs in the region chr16:0.65-2.02Mb of one normal (dark gray) and tumor samples (light gray) beloning to BRCA dataset from TCGA.

5.3 Single CNV association

Genome-wide association analysis of CNVs aims to determine whether *any* CNV in the population sample is associated with a phenotype of interest. However, targeted analyses are also important when a CNV has been hypothesized. One of the key issues when assessing such association is how to account for the uncertainty derived from CNV calling in the association models [41].

We illustrate a single CNV association analysis with data that includes 360 cases and 291 controls [41]. The data contains the peak intensities for two genes arising from an MLPA assay. Note that for Illumina or Affymetrix SNP array CNV status will be given by LRR instead of peak intensities. Data can be loaded into R by:

```
> data(dataMLPA, package="CNVassocData")
> head(dataMLPA)
    id casco Gene1     Gene2 PCR.Gene1 PCR.Gene2 quanti    cov
1 H238     1  0.51 0.5385080        wt        wt  -0.61  10.83
2 H238     1  0.45 0.6392029        wt        wt  -0.13  10.69
3 H239     1  0.00 0.4831572       del        wt  -0.57   9.63
4 H239     1  0.00 0.4640072       del        wt  -1.40   9.87
5 H276     1  0.00 0.0000000       del       del   0.83  10.25
6 H276     1  0.00 0.0000000       del       del  -2.07  10.40
```

The MLPA dataset contains case-control status as well as two simulated covariates (`quanti` and `cov`) that have been generated for illustrative purposes. We perform the association analysis with the *CNVassoc* package which fits a likelihood model that incorporates the uncertainty of CNV calling

```
> library(CNVassoc)
```

We first visualize the distribution of peak intensities of two genes in the data (Figure 5.5) that can be reproduce by executing:

```
> par(mfrow=c(2,2),mar=c(3,4,3,1))
> hist(dataMLPA$Gene1,main="Gene 1 signal histogram",xlab="",ylab="frequency")
> hist(dataMLPA$Gene2,main="Gene 2 signal histogram",xlab="",ylab="frequency")
> par(xaxs="i")
> plot(density(dataMLPA$Gene1), main="Gene 1 signal density function",xlab="",
+      ylab="density")
> plot(density(dataMLPA$Gene2), main="Gene 2 signal density function",xlab="",
+      ylab="density")
```

The figure shows the signals for Gene 1 and Gene 2 of the MLPA data example. For both genes, it is clear that there are 3 clusters corresponding to 0, 1 and 2 copies. However, the three peaks for Gene 2 are not so well separated as those of Gene 1 (the underlying distributions overlap much more). Therefore CNV calling for Gene 2 carries large uncertainty.

The function `plotSignal` can be used to describe the peak intensities of Gene 2 in cases and controls (Figure 5.6)

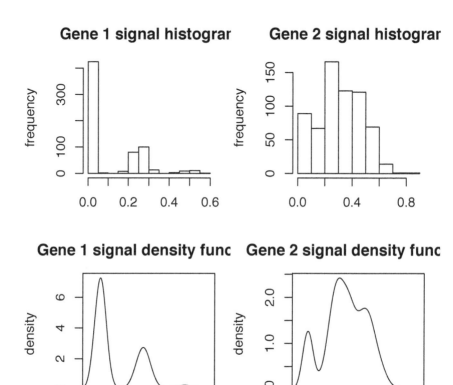

FIGURE 5.5
Signal distributions for Gene 1 and Gene 2 from MLPA data example.

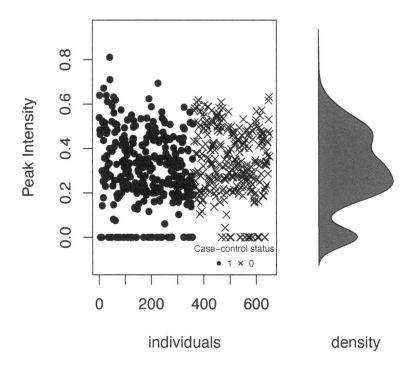

FIGURE 5.6
Signal distribution for cases and controls of Gene 2 from MLPA data example.

```
> with(dataMLPA, plotSignal(Gene2, case.control=casco))
```

A similar plot can be obtained for a quantitative trait

```
> with(dataMLPA, plotSignal(Gene2, case.control = quanti))
```

In Figure 5.7, the quantitative phenotype is plotted on the x-axis, instead of distinguishing points by shape, as in Figure 5.6.

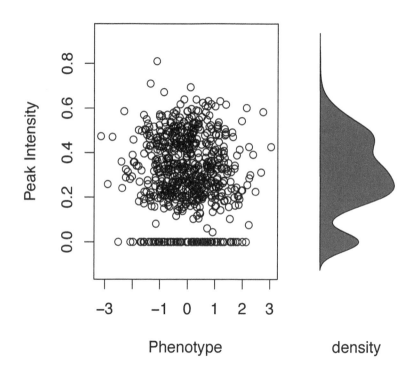

FIGURE 5.7
Signal distribution of Gene 2 from MLPA data example for a quantitative trait.

5.3.1 Inferring copy number status from signal data

The main calling algorithm in `cnv` of *CNVassoc* is based on a normal mixture model [109, 162]. Note that in some cases, the intensity distributions (see Gene 1 in Figure 5.5) for a null allele are expected to be equal to 0. Due to experimental noise, these intensities can deviate slightly from this theoretical value. For these cases, the normal mixture model fails because the underlying distribution of individuals with 0 copies is not normal. In these situations, we can fit a modified mixture model (see [41] for further details). We apply `cnv` for Gene 1 in figure 5.5

```
> CNV.1 <- cnv(x = dataMLPA$Gene1, threshold.0 = 0.06,
+              num.class = 3, mix.method = "mixdist")
```

With argument `threshold.0=0.06`, individuals with gene intensities lower than 0.06 will be called with 0 copies for Gene 1. We see that there are three underlying status of copy number for individuals thus we set argument `num.class` to 3. With the argument `mix.method` we select the algorithm to estimate the normal mixture model. `"mixdist"` uses the EM algorithm implemented in the *mixdist* package while `"mclust"` uses the EM implemented in the *Mclust* package. The generic `print` function summarizes the inferred copy number status

```
> CNV.1

Inferred copy number variant by a quantitative signal
   Method: function mix {package: mixdist}

-. Number of individuals: 651
-. Copies 0, 1, 2
-. Estimated means: 0, 0.2543, 0.4958
-. Estimated variances: 0, 9e-04, 0.0012
-. Estimated proportions: 0.6544, 0.3088, 0.0369
-. Goodness-of-fit test: p-value= 0.6615318
```

It displays the means, variances and proportions of copy number clusters as well as the *P*-value corresponding to the goodness-of-fit for the selected number of classes. `CNV.1` is an object of class *cnv* that can be plotted to visualize the CNV calling

```
> plot(CNV.1, case.control = dataMLPA$casco, main = "Gene 1")
```

As shown in figures 5.8, the signal is colored by the most probable copy number while cases and controls are distinguished by shape, specified by the argument `case.control`. On the right side of the plot, a density function of signal distribution is drawn. The *P*-value of goodness-of-fit is also given for the normal mixture model.

When the exact number of components for the mixture model is not known, like the case of Gene 2, `cnv` uses the Bayesian Information Criteria (BIC) to

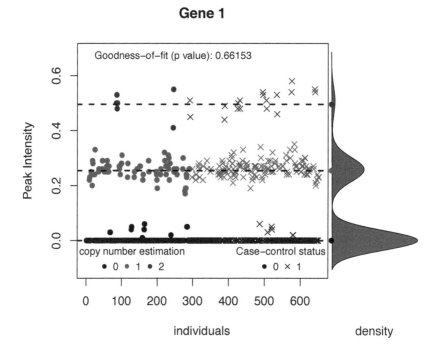

FIGURE 5.8
CNV calling of Gene 1 from the MLPA example.

select the best number of components. This is performed when the argument
num.class is missing. In this case, the function estimates the mixture model
considering from 2 up to 6 copy number status and selects the best one. In the
case of Gene 2 the function selects 3 classes as the best model for the data,
as expected from experimental results [41].

```
> CNV.2 <- cnv(x = dataMLPA$Gene2, threshold.0 = 0.01,
+               mix.method = "mixdist")
> CNV.2

Inferred copy number variant by a quantitative signal
   Method: function mix {package: mixdist}

-. Number of individuals: 651
-. Copies 0, 1, 2
-. Estimated means: 0, 0.2435, 0.4469
-. Estimated variances: 0, 0.0041, 0.0095
-. Estimated proportions: 0.1306, 0.4187, 0.4507
-. Goodness-of-fit test: p-value= 0.4887659


-. Note: number of classes has been selected using the best BIC
```

The CNV calling of Gene 2 is shown in Figure 5.9

```
> plot(CNV.2, case.control = dataMLPA$casco, main = "Gene 2")
```

The uncertainty in the CNV calling is saved as an attribute of *cnv*, and
retrieved by the **getProbs** function

```
> probs.2 <- getProbs(CNV.2)
> head(probs.2)
      [,1]          [,2]        [,3]
[1,]     0 5.552007e-05 0.9999445
[2,]     0 5.293850e-08 0.9999999
[3,]     0 1.398147e-03 0.9986019
[4,]     0 3.860874e-03 0.9961391
[5,]     1 0.000000e+00 0.0000000
[6,]     1 0.000000e+00 0.0000000
```

showing the posterior probabilities of CNV status for each individual. These
probabilities are used in the association analysis to account for uncertainty in
CNV calling. If an uncertainty matrix **probs.2** was obtained by other calling
algorithms, such as CANARY (from PLINK) or **GCHcall**, we can perform an
association analysis on those estimates by creating an object of class *cnv* with
the uncertainty matrix

```
> CNV.2probs <- cnv(probs.2)
```

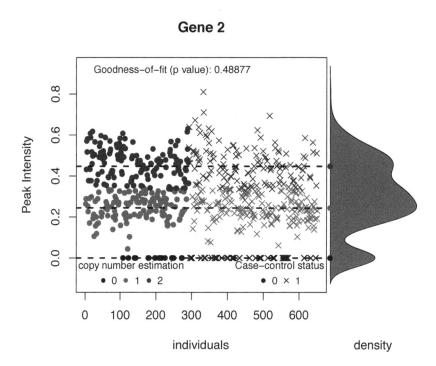

FIGURE 5.9
CNV calling of Gene 2 from the MLPA example.

5.3.2 Measuring uncertainty of CNV calling

The function `getQualityScore` measures the degree of overlapping in the copy number mixture distribution. The more separated these peaks are, the less uncertainty there is, and the larger the measure is. Three quality measures on the overall uncertainty call are currently implemented that can be passed through an object of class *cnv*. The first one used is the one in the *CNVtools* package, the second one is the estimated probability of good classification (PGC), and the third one (CANARY) is defined as the proportion of individuals with a confidence score bigger than 0.1 [72]. They are illustrated as follows. The measure defined in the *CNVtools* package can be obtained by

```
> CNVassoc::getQualityScore(CNV.1, type = "CNVtools")
--CNVtools Quality Score: 25.16849
> CNVassoc::getQualityScore(CNV.2, type = "CNVtools")
--CNVtools Quality Score: 3.057171
```

The PGC is computed with the argument `type` set to `"class"`

```
> CNVassoc::getQualityScore(CNV.1, type = "class")
--Probability of good classification: 0.9999779
> CNVassoc::getQualityScore(CNV.2, type = "class")
--Probability of good classification: 0.9114605
```

and the third measure is obtained by

```
> CNVassoc::getQualityScore(CNV.1, type = "CANARY")
--Probability to have a 'CANARY confidence index' > 0.1 : 0
> CNVassoc::getQualityScore(CNV.2, type = "CANARY")
--Probability to have a 'CANARY confidence index' > 0.1 : 0.3024652
```

The measures show that Gene 1 carries much less uncertainty; PGC is greater than 99%, the *CNVtools* score is higher than 25 (recommended quality > 4), and CANARY score is close to 0. We can appreciate this low uncertainty in figure 5.8, where the underlying distributions of signal intensity are very well separated. On the other hand, for Gene 2 we have a different picture, more uncertainty is present (Figure5.9); PGC is 91.3%, the *CNVtools* score value is 3 and the CANARY measure is 30%.

When a *cnv* object is created from probabilities of other calling algorithms only the CANARY measure is computed, for which CNVs with values > 0.1 or 10% should not be analyzed [99].

5.3.3 Assessing the association between CNVs and traits

The function `CNVassoc` carries out association analysis between CNV and disease. This function incorporates calling uncertainty by using a latent class model as described in [41]. The function can analyze both binary and quantitative traits. In the first case, a linear regression is performed, and, in the second, a logistic regression. The regression model can be selected by using the argument `case.control`. Nonetheless, the program automatically detects whether a quantitative trait is analyzed so that, this argument is not required to be specified.

The function also allows the user to fit a model with additive or multiplicative effects of CNV. This can be set through the argument `model`. Possible values are "add for an additive effect or "mul for a multiplicative effect. The function CNVassoc returns an object of class *CNVassoc*. This class of object has implemented generic methods similar to the ones of objects of class *glm*, such as `coef` or `summary` among others.

5.3.3.1   Modeling association

The effect of a given CNV on case/control status (`casco` variable in the MLPA example) can be fitted by typing:

```
> model1mul <- CNVassoc(casco ~ CNV.1, data = dataMLPA, model = "mul")
> model2mul <- CNVassoc(casco ~ CNV.2, data = dataMLPA, model = "mul")
```

The results of these analyses are similar to the ones obtained from `glm` function that implements generalized linear models.

```
> model1mul

Call:  CNVassoc(formula = casco ~ CNV.1, data = dataMLPA, model = "mul")

Coefficients:
          CNV0        CNV1       CNV2
CNVmult  0.0281709  0.5187566  1.0989109

Number of individuals: 651
Number of estimated parameters: 3
Deviance: 883.03
```

Note that the coefficients are a matrix with one row per variable and a column for each distinct copy number statuses. In this model, because there are no covariates and the CNV has a multiplicative effect, there is just one row (one intercept) and this is different among columns (copy number status).

By using the generic function `summary`, we can obtain a more exhaustive output. In particular, the odds ratio and its confidence intervals are printed as well as its *P*-value.

```
> summary(model1mul)

Call:
CNVassoc(formula = casco ~ CNV.1, data = dataMLPA, model = "mul")

Deviance: 883.0297
Number of parameters: 3
Number of individuals: 651

Coefficients:
```

```
          OR lower.lim upper.lim     SE   stat pvalue
CNV0 1.0000
CNV1 1.6333     1.1588     2.3020 0.1751 2.8017  0.005
CNV2 2.9175     1.1359     7.4937 0.4813 2.2247  0.026


(Dispersion parameter for  binomial  family taken to be  1 )


Covariance between coefficients:
      CNV0    CNV1    CNV2
CNV0 0.0094 0.0000 0.0000
CNV1        0.0213 0.0000
CNV2               0.2223
> summary(model2mul)

Call:
CNVassoc(formula = casco ~ CNV.2, data = dataMLPA, model = "mul")

Deviance: 876.396
Number of parameters: 3
Number of individuals: 651

Coefficients:
          OR lower.lim upper.lim     SE    stat pvalue
CNV0   1.0000
CNV1   0.4772    0.2742    0.8304 0.2827 -2.6172  0.009
CNV2   0.3169    0.1834    0.5477 0.2791 -4.1169  0.000

(Dispersion parameter for  binomial  family taken to be  1 )


Covariance between coefficients:
     CNV0    CNV1    CNV2
CNV0  0.0613  0.0000  0.0000
CNV1          0.0186 -0.0032
CNV2                  0.0166
```

We can see, in particular, that CNV gains of Gene 1 significantly increase the risk of being a case while CNV gains of Gene 2 significantly reduce the risk.

By default, CNVassoc treats the response variable as a binary phenotype coded as 0/1. Since CNVassoc can handle other distributions such as Poisson or Weibulls, the family argument must be specified when the response is not distributed as a Bernoulli distribution. For instance, to deal with a normally distributed response variable, specify family="gaussian".

The following example presents the case of analyzing a quantitative normally distributed trait and adjusting the association by other covariates:

```
> mod <- CNVassoc(quanti ~ CNV.2 + cov, family = "gaussian",
+                 data = dataMLPA, model = 'add', emsteps = 10)
> coefficients(mod)
                CNV0          CNV1          CNV2
intercept -0.14037609 -0.14037609 -0.14037609
CNVadd    -0.07923667 -0.07923667 -0.07923667
```

```
cov        0.02418770  0.02418770  0.02418770
```

Notice that in this case, we use the argument `emsteps`. This is necessary for computational reasons. Initially performing some preliminary steps using the EM algorithm makes it easier to maximize the likelihood function using the Newton–Raphson procedure. In general, it is enough to perform a few iterations (no more than 10). As usual, the model is then summarized by typing:

```
> summary(mod)

Call:
CNVassoc(formula = quanti ~ CNV.2 + cov, data = dataMLPA, model = "add",
    family = "gaussian", emsteps = 10)

Deviance: 1824.573
Number of parameters: 4
Number of individuals: 651

Coefficients:
              beta lower.lim upper.lim      SE     stat pvalue
(Intercept) -0.14038  -0.90687   0.62612 0.39108 -0.35895  0.720
trend       -0.07924  -0.19714   0.03866 0.06015 -1.31722  0.188
cov          0.02419  -0.05068   0.09906 0.03820  0.63321  0.527

(Dispersion parameter estimation for  gaussian family is 0.9650261 )


Covariance between coefficients:
          intercept CNVadd   cov
intercept   0.1529   -0.0041 -0.0146
CNVadd                0.0036 -0.0001
cov                          0.0015
```

Remember that for quantitative traits we obtain mean differences instead of odds ratios.

5.3.3.2   Global test of associations

In the previous analysis, we obtained P-values corresponding to the comparison between every copy number status versus the reference (zero copies). Nonetheless, we are normally interested in testing the overall effect of CNV on a disease. The Wald test and the likelihood ratio test (LRT) are implemented to perform such omnibus testing. Both are available through the function `CNVtest` which requires an object of class `CNVassoc` as the input. To specify the type of test, set the argument `type` to `"Wald"` or `"LRT"`, respectively. For Gene 1,

```
> CNVtest(model1mul, type = "Wald")
----CNV Wald test----
Chi= 11.55332 (df= 2 ) , pvalue= 0.003099052
> CNVtest(model1mul, type = "LRT")
```

```
----CNV Likelihood Ratio Test----
Chi= 12.12081 (df= 2 ) , pvalue= 0.002333458
```

Other generic functions like `logLik`, `coef`, `summary` or `update` can be applied to an object of class `CNVassoc` to get more information.

For a multiplicative CNV effect model and for a binary traits, it is possible to change the reference category of copy number status. This can be done by using the argument `ref` when executing the `summary` function. For example, if we want one copy as the reference category just type:

```
> coef(summary(model1mul, ref = 2))
            OR lower.lim upper.lim        SE      stat     pvalue
CNV1 1.0000000        NA        NA        NA        NA         NA
CNV0 0.6122677 0.4344016 0.8629612 0.1751053 -2.801661 0.005084028
CNV2 1.7863141 0.6790498 4.6990929 0.4934862  1.175624 0.239745081
```

The same kind of results can be obtained if we assume an additive effect of CNV on the trait. In this case we need to set the `model` argument to `"add"`

```
> model2add <- CNVassoc(casco ~ CNV.2, data = dataMLPA, model = "add")
> model2add

Call:  CNVassoc(formula = casco ~ CNV.2, data = dataMLPA, model = "add")

Coefficients:
              CNV0        CNV1        CNV2
intercept  0.932028    0.932028    0.932028
CNVadd    -0.537731   -0.537731   -0.537731

Number of individuals: 651
Number of estimated parameters: 2
Deviance: 877.061
```

Notice that under an additive CNV effect the structure of coefficients is different from the multiplicative CNV effect. Now there are two rows, one for intercept and the other one for the slope (increment of risk by one copy). These two values remain constant for every column (copy number status).

```
> summary(model2add)

Call:
CNVassoc(formula = casco ~ CNV.2, data = dataMLPA, model = "add")

Deviance: 877.0606
Number of parameters: 2
Number of individuals: 651

Coefficients:
           OR lower.lim upper.lim     SE     stat pvalue
trend  0.5841    0.4530    0.7530 0.1296 -4.1477      0

(Dispersion parameter for  binomial  family taken to be  1 )
```

```
Covariance between coefficients:
          intercept CNVadd
intercept  0.0374   -0.0228
CNVadd               0.0168
```

Finally, one might be interested in testing the additive effect. To do this, one can compare both additive and multiplicative models. It is straightforward to see that the additive model is a particular case of the multiplicative one, and therefore the first is nested in the second one. To compare two nested models we use the generic function **anova** (NOTE: it is only implemented for comparing two models, both fitted with the **CNVassoc** function).

```
> anova(model2mul, model2add)


--- Likelihood ratio test comparing 2 CNVassoc models:


Model 1 call: CNVassoc(formula = casco ~ CNV.2, data = dataMLPA, model = "mul")

Model 2 call: CNVassoc(formula = casco ~ CNV.2, data = dataMLPA, model = "add")

Chi= 0.6645798 (df= 1 )  p-value= 0.4149477

  Note: the 2 models must be nested, and this function doesn't check this!
```

The likelihood ratio test is performed. In this case the *P*-value is not significant, indicating that an additive CNV effect can be assumed. In any case, one should consider the power of this test before making conclusions.

5.3.4   Whole genome CNV analysis

The analysis of aCGH data requires taking additional steps into account since CNV status is measured in different probes and this produces dependency across them. Table 5.3.4 shows four steps we recommend for the analysis of this kind of data. First, posterior probabilities should be obtained with an algorithm that considers probe correlation. We use, in particular, the *CGHcall* R package which includes a mixture model to infer CNV status [162]. Second, we build blocks/regions of consecutive probes with similar signatures. To perform this step the *CGHregions* R package is recommended [163]. Third, the association between the CNV status of blocks and the trait is assessed by incorporating the uncertainty probabilities in the **CNVassoc** function. And fourth, corrections for multiple comparisons must be performed. We use the Benjamini–Hochberg (BH) correction [10], a widely used method for control of FDR that is robust in the scenarios commonly found in genomic data [121].

To illustrate, we apply these steps to the breast cancer data studied by Neve et al. [102]. The data consists of CGH arrays of 1MB resolution and is available from Bioconductor http://www.bioconductor.org/. The authors

TABLE 5.2
Steps to assess association between CNVs and traits for aCGH.

Step 1. Use any CNV calling procedure that provides posterior probabilities (uncertainty) (*CGHcall*)
Step 2. Build blocks/regions of consecutive probes with similar signatures (*CGHregions*)
Step 3. Use the signature that occurs most in a block to perform association (**multiCNVassoc** from *SNPassoc*)
Step 4. Correct for multiple testing considering dependency among signatures (**getPvalBH** from *CNVassoc*)

chose the 50 samples that could be matched to the name tokens of caArrayDB data (June 9th, 2007). In this example, the associations between estrogen receptor positivity (dichotomous variable; 0: negative, 1: positive) and CNVs was tested. The data is saved in an object called **NeveData** in the *brgedata*. This object is a list with two components. The first component corresponds to a *data.frame* containing 2,621 rows and 54 columns with aCGH data (4 columns for the annotation and 50 log2ratio intensities). The second component is a vector with the phenotype analyzed (estrogen receptor positivity). The data can be loaded as usual

```
> data(NeveData, package="CNVassocData")
> intensities <- NeveData$data
> pheno <- NeveData$pheno
```

The original data set contained 2,621 probes which were reduced to 459 blocks after the application of *CGHcall* and *CGHregions* as following:

```
>    #####################################################
>    ### chunk number 1: Class of aCGH data
>    #####################################################
>    library(CGHcall)
>    Neve <- make_cghRaw(intensities)
>
>    #####################################################
>    ### chunk number 2: Preprocessing
>    #####################################################
>    cghdata <- preprocess(Neve, maxmiss = 30, nchrom = 22)
>
>    #####################################################
>    ### chunk number 3: Normalization
>    #####################################################
>    norm.cghdata <- normalize(cghdata, method = "median",
+                              smoothOutliers = TRUE)
>
>    #####################################################
>    ### chunk number 4: Segmentation
>    #####################################################
```

```
>    seg.cghdata <- segmentData(norm.cghdata, method = "DNAcopy")
>
>    ####################################################
>    ### chunk number 5: Calling
>    ####################################################
>    NeveCalled <- CGHcall(seg.cghdata, nclass = 3)
>    NeveCalled <- ExpandCGHcall(NeveCalled, seg.cghdata)
```

We can then obtain the posterior probabilities. `CGHcall` function does not estimate the underlying number of copies for each segment but assigns the underlying statuses: loss, normal or gain. For each segment and for each individual, we obtain three posterior probabilities corresponding to each of these three statuses. This is done by executing

```
> probs <- getProbs(NeveCalled)
> probs[1:5, 1:7]
               Clone Chromo BPstart    BPend X600MPE X600MPE.1 X600MPE.2
RP11-82D16 RP11-82D16      1 2008651 2008651       0     0.973     0.027
RP11-62M23 RP11-62M23      1 3367844 3367844       0     0.973     0.027
RP11-11I05 RP11-11I05      1 4261844 4261844       0     0.973     0.027
RMC01P070   RMC01P070      1 5918606 5918606       0     0.973     0.027
RP11-51B4   RP11-51B4      1 6068980 6068980       0     0.973     0.027
```

This table can be read as follows. The probabilities that the individual `X600MOE` has a losss, is normal and has a gain for the signature `RP11-82D16` are 0, 0.93 and 0.027, respectively. In order to determine the regions that are recurrent or common among samples, we use the *CGHregions* function that takes an object of class *cghCall* (e.g. object `NeveCalled` in our case). This algorithm reduces the initial table to a smaller matrix that contains regions rather than individual probes. The regions consist of consecutive clones with similar signatures [163]. This can be done by executing:

```
> library(CGHregions)
> NeveRegions <- CGHregions(NeveCalled)
```

Getting the posterior probabilities for each block/region to be used in the association analysis incorporating uncertainty can be done by:

```
> probsRegions <- getProbsRegions(probs, NeveRegions, intensities)
```

Lastly, the association analysis between each region and the estrogen receptor positivity can be analyzed by using the `multiCNVassoc` function. This function repeatedly calls `CNVassoc` returning the *P*-value of association for each block/region

```
> pvalsCNV <- multiCNVassoc(probsRegions, formula = "pheno ~ CNV",
+                           model = "mult", num.copies = 0:2,
+                           cnv.tol = 0.01)
```

Notice that the arguments of `multiCNVassoc` function are the same as those of `CNVassoc`. In this example, we have set the argument `num.copies` equal to 0, 1, and 2 that corresponds to `loss, normal, gain` statuses used in the `CGHcall` function.

Multiple comparisons can be addressed by using the Benjamini–Hochberg approach [10]. The function `getPvalBH` produces the FDR-adjusted P-values

```
> pvalsBH <- getPvalBH(pvalsCNV)
> head(pvalsBH)
  region       pval      pval.BH
1    274 7.005210e-06 0.002746042
2    257 7.904878e-05 0.006528060
3    273 5.085502e-05 0.006528060
4    275 5.496328e-05 0.006528060
5    277 8.326607e-05 0.006528060
6    363 1.446898e-04 0.009453070
```

Table 6 in [41] can be reproduced by:

```
> cumsum(table(cut(pvalsBH[, 2], c(-Inf, 1e-5, 1e-4, 1e-3, 1e-2, 0.05))))
   (-Inf,1e-05] (1e-05,0.0001] (0.0001,0.001]    (0.001,0.01]   (0.01,0.05]
              1              5             23              50            97
```

5.4 Genetic mosaicisms

The detection of genomic mosaicism, the coexistence of cells of distinct genetic composition within an organism, can be performed by any method that allows observing cellular differences in the same biological sample. This includes from classic karyotyping analysis to fluorescence in situ hybridization [132], comparative-genomic hybridization and its array variation [136], PCR-based methods (qPCR, DD-PCR), protein truncation test, or even sequencing methods (whole-genome, whole-exome, targeted sequencing, single cell, and molecular inversion probes among others). Depending on its resolution, each technique is able to detect different types of genomic mosaic events. For instance, SNP arrays can be used to detect chromosomal mosaicisms, as they are based on a fixed set of variants.

5.4.1 Calling genetic mosaicisms

The *MAD* package was developed to deect chromosomal mosaicism using SNP array data [45]. The method is based on the detection of consecutive SNP positions with altered BAF using the GADA methodology [112]. Deviations from the expectations indicate a copy-number change, usually in the form of the separation of the BAF of heterozygous SNPs from its expected value 0.5. For instance, if a chromosomal region is duplicated for all of the cells of an individual's biological sample, a heterozygous SNP within the region will show double the amount of allelic signal from the chromosome with the duplication than from the normal one. However, if the duplication is for half of the cells then only the signal increment will be halved, showing a signal for the duplication that is three-fourths of the normal one. In these cases, the BAF will be displaced up or below 0.5 depending on the reference allele and the chromosome with the duplication and the number of cells affected by it. Contiguous heterozygous SNPs within the duplication will show a similar pattern, some above and some below 0.5, but displaced at the same amount, in what is known as the BAF split that extends the region with the copy number change. The distance between the two bands from 0.5 is known as B-deviation (Bdev) [126]. Segmentation of the Bdev split in heterozygous SNPs is then used to detect chromosome regions affected by mosaicisms. These regions are then classified according to LRR values between gains (LRR > 0), losses (LRR < 0) and copy-neutral loss of heterozygosity (Uniparental disomy) events (LRR ~ 0). The analysis of genetic mosaicisms using SNP array data has led to the discovery of their relationship to diseases, [18] including its association with aging and cancer [64], Fanconi anemia [120] or developmental disorders [70].

We illustrate the use of *MAD*

```
> devtools::install_github("isglobal-brge/MAD")
```

The package is loaded as usual

```
> library(mad)
```

As in the case of CNV calling with *R-GADA* (see Section 5.2.1), *MAD* requires data in PennCNV format and they are located in a folder called **rawData**. We show how to perform mosaic calling using the data in *brgedata* package. We load the data from the package and configure the files that are needed for mosaic call

```
> ss1 <- system.file("extdata/madData", package="brgedata")
> dir.create("rawData")
> ss2 <- "rawData"
> files <- list.files(ss1)
> file.copy(file.path(ss1,files), ss2)
```

The analysis starts by computing the Bdev of each sample:

```
> mosaic <- setupParGADA.B.deviation(NumCols=6, GenoCol=6,
+                                    BAFcol=5, log2ratioCol=4)

Creating object with annotation data ...
Creating object with annotation data ...done

Creating objects of class setupGADA for all input files...
  Applying setupGADA.B.deviation for 5 samples ...
```

```
Importing array:  CASE369.txt ...    Array # 1 ...done
Importing array:  CASE371.txt ...    Array # 2 ...done
Importing array:  CASE377.txt ...    Array # 3 ...done
Importing array:  CONTROL152.txt ...   Array # 4 ...done
Importing array:  CONTROL191.txt ...   Array # 5 ...done
 Applying setupGADA.B.deviation for 5 samples ... done
Creating objects of class setupGADA for all input files... done
```

where we specify in `GenoCol`, `BAFcol` and `log2ratioCol` arguments, the columns in the files where to find the genotypes, BAF and LRR. The first three columns of the PennCNV must be snp, chromosome and position, in that order. The object `mosaic` contains the following information:

```
> mosaic
[1] "C:/Juan/CREAL/GitHub/multiomic_book/data"
attr(,"class")
[1] "parGADA"
attr(,"labels.samples")
[1] "CASE369"    "CASE371"    "CASE377"    "CONTROL152" "CONTROL191"
attr(,"Samples")
[1] 5
attr(,"b.deviation")
[1] TRUE
```

Segmentation on Bdev is performed using the two steps implemented *R-GADA* previously illustrated in the case of CNV calling (Section 5.2.1):

```
> parSBL(mosaic, estim.sigma2=TRUE, aAlpha=0.8)
Creating SBL directory ...done
Retrieving annotation data ...done
Segmentation procedure for 5 samples ...
   Array # 1 ...    The estimated sigma2 = 0.910931
   Array # 1 ...done
   Array # 2 ...    The estimated sigma2 = 0.754135
   Array # 2 ...done
   Array # 3 ...    The estimated sigma2 = 1.043872
   Array # 3 ...done
   Array # 4 ...    The estimated sigma2 = 0.9110874
   Array # 4 ...done
   Array # 5 ...    The estimated sigma2 = 0.371038
   Array # 5 ...done
Segmentation procedure for 5 samples ...done
> parBE.B.deviation(mosaic, T=9, MinSegLen=100)
Retrieving annotation data ...done
Backward elimination procedure for 5 samples ...
   Array # 1 ... -------------------------------------
Sparse Bayesian Learnig (SBL) algorithm
Backward Elimination procedure with T=9 and minimun length size=100
 Number of segments =  53
 Base Amplitude of copy number 2: chr 1:22:0.558, X=0.0167, Y=0
   Array # 2 ... -------------------------------------
Sparse Bayesian Learnig (SBL) algorithm
Backward Elimination procedure with T=9 and minimun length size=100
 Number of segments =   43
```

```
Base Amplitude of copy number 2: chr 1:22:0.3779, X=0.0226, Y=0
  Array # 3 ... ---------------------------------------
Sparse Bayesian Learnig (SBL) algorithm
Backward Elimination procedure with T=9 and minimun length size=100
 Number of segments =  68
 Base Amplitude of copy number 2: chr 1:22:0.705, X=0.0517, Y=0
  Array # 4 ... ---------------------------------------
Sparse Bayesian Learnig (SBL) algorithm
Backward Elimination procedure with T=9 and minimun length size=100
 Number of segments =  66
 Base Amplitude of copy number 2: chr 1:22:0.6464, X=0.0431, Y=0
  Array # 5 ... ---------------------------------------
Sparse Bayesian Learnig (SBL) algorithm
Backward Elimination procedure with T=9 and minimun length size=100
 Number of segments =  35
 Base Amplitude of copy number 2: chr 1:22:0.2624, X=0.0028, Y=0
Backward elimination procedure for 5 samples ...done
```

The parameter **aAlpha** controls the number of breakpoints. Following the recommendations in Table 5.1, we set **aAlpha=0.8** for Illumina 1M. The parameter T controls the False Discovery Rate (FDR) and is increased to decrease FDR. The user can change the parameter T to provide a longer or shorter list of mosaic regions. The argument **MinSegLen** indicates the minimum number of consecutive heterozygous SNPs that have Bdev different from 0, setting the minimum size of detectable mosaic events. The detected mosaic regions are

```
> mosaic.gr <- getMosaics(mosaic)
```

containing the following information

```
> mosaic.gr
GRanges object with 42 ranges and 6 metadata columns:
        seqnames                ranges strand | LenProbe       LRR      LRR.se
           <Rle>             <IRanges>  <Rle> | <integer> <numeric> <numeric>
   [1]     chr6 31578652-32298368       * |      1522         0      0.14
   [2]     chr8 41574161-49215718       * |      1266      0.01      0.25
   [3]     chr9 130037713-131048917     * |       525      0.02      0.15
   [4]    chr14 65674121-66989436       * |       477      0.04      0.17
   [5]    chr15 40160724-43767684       * |      1599      0.02      0.19
   ...      ...                   ...    ... .       ...       ...       ...
  [38]    chr16 31688557-47128764       * |      1186     -0.05      0.29
  [39]    chr16 64931218-66531454       * |       796     -0.03      0.21
  [40]     chrX 814-2663913             * |       501     -0.14      0.32
  [41]    chr20 14405834-44638545       * |     11501     -0.12      0.29
  [42]    chr20 44638889-48596642       * |      1763     -0.22      0.29
            Bdev     State      sample
        <numeric> <numeric> <character>
   [1]     0.243         5     CASE369
   [2]     0.315         5     CASE369
   [3]     0.204         5     CASE369
   [4]     0.323         5     CASE369
   [5]     0.245         5     CASE369
   ...       ...       ...         ...
  [38]     0.354         1  CONTROL152
```

```
[39]      0.32       1   CONTROL152
[40]      0.314      2   CONTROL152
[41]      0.108      2   CONTROL191
[42]      0.199      2   CONTROL191
-------
seqinfo: 16 sequences from an unspecified genome; no seqlengths
```

The column **State** shows the classification of mosaic alterations based on the segmentation of B-deviation. The number codes correspond to the following abnormalities: Uniparental disomy (1), deletion (2), duplication (3), trisomy (4) and LOH (5). It is recommendable to visually inspect detected mosaics by plotting results with **plotQMosaic**, which shows the LRR and BAF values, see figure 5.10 that corresponds to the sample CONTROL191 that has mosaic chromosome 20

```
> # argument 'col' is not necessary
> mosaic <- addAnnot(mosaic) # add annotation
> plotQMosaic(mosaic, sample="CONTROL191", chr=20,
+             regions=mosaic.gr,
+             col.dots=c("black", "gray50"))
```

We can zoom out the plot and depict the altered region for better visualization (Figure 5.11)

```
> plotZoomQMosaic(mosaic, sample="CONTROL191", chr=20,
+                 regions = mosaic.gr,
+                 col.dots=c("black", "gray50"))
```

Mosaic segments at individual level for the whole genome can be plotted using *GenVisR* Bioconductor package. These are depicted in the Figure 5.12 that can be created by plotting the segments of the object containing **GenomicRanges** of mosaic alterations (e.g. **mosaic.gr**):

```
> library(ggplot2)
> plotSegments(mosaic.gr)
```

We can also visualize any region of interest by using the function **plotCNVs** from *R-GADA* by setting **mosaic=TRUE** (Figure 5.13)

```
> rr <- GRanges("chr16:63.5e6-67.0e6")
> plotCNVs(mosaic.gr, range=rr, drawGenes = TRUE, mosaic=TRUE)
```

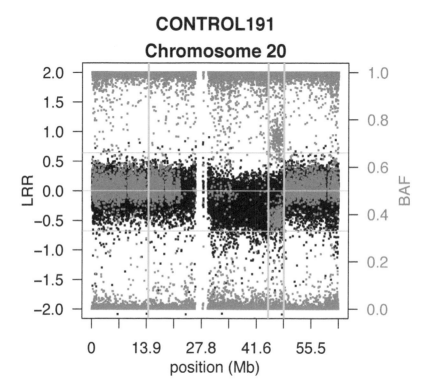

FIGURE 5.10
LRR and BAF corresponding to sample CONTROL191 from *brgedata* which
was detected as mosaic loss.

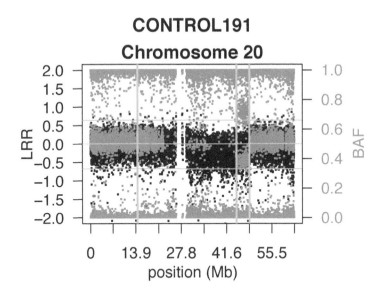

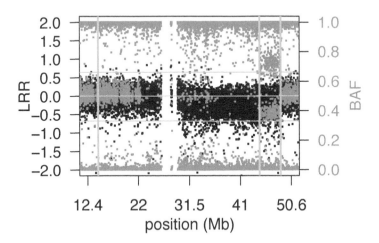

FIGURE 5.11
LRR and BAF corresponding to sample CONTROL191 from *brgedata* detected as mosaic loss having the altered region delimited by vertical lines.

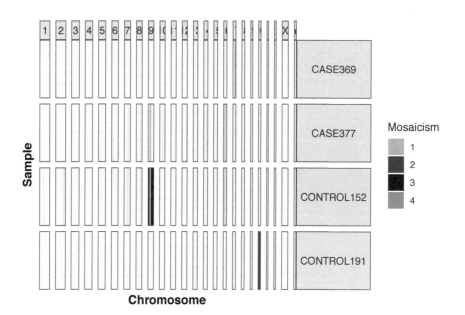

FIGURE 5.12
Copy number alterations at genome level for different samples from *brgedata*.
The legend corresponds to: UPD (1), deletion (2), duplication (3), trisomy
(4).

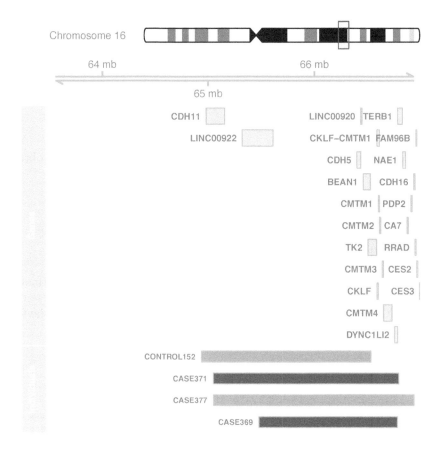

FIGURE 5.13
Mosaic alterations UPD (light gray) and LOH (dark gray) in the region chr8:63.5-67Mb of samples belonging to data available at the *brgedata* package.

5.4.2   Calling the loss of chromosome Y

Mosaic loss of chromosome Y (LOY) is a particular case of genetic mosaicism that has been associated with an increased risk of cancer[36], Alzheimer's disease [32] and major cardiovascular defects [49]. The calling of LOY status can be performed from LRR values across all SNPs probes in the male-specific region of chromosome Y (mLRR-Y), comprising the 56-Mb region between pseudoautosomal regions 1 and 2 (PAR1 and PAR2) (chrY:2,694,521-59,034,049, hg19/GRCh37) [36]. We demonstrate LOY calling from LRR data using *MADloy* package that can be installed from GitHub as:

```
> devtools::install_github("isglobal-brge/MADloy")
```

and loaded as usual

```
> library(MADloy)
```

We analyze data including 124 males and 2 females that can be obtained from our *brgedata* package.

```
> ss1 <- system.file("extdata/rawData", package="brgedata")
> dir.create("LOY/rawData")
> ss2 <- "LOY/rawData"
> files <- list.files(ss1)
> file.copy(file.path(ss1,files), ss2)
```

The data include one file per sample in the required PennCNV format. We perform the calling of three possible status of Y mosaicism in the samples: LOY/XYY/normal. We decompress the .zip file in a folder and set the folder path in `rawDataPath`

```
> rawDataPath <- "LOY/rawData"
> files <- dir(rawDataPath)
> length(files)
[1] 126
> files[1:5]
[1] "FEMALE_2291.txt" "FEMALE_2439.txt" "SAMPLE_1.txt"   "SAMPLE_10.txt"
[5] "SAMPLE_100.txt"
```

LOY is a male-specific mosaicism. If females are present in the data, the function *checkSex* can detect them to remove them from the analysis

```
> sex <- checkSex(rawDataPath, LRRCol=4)
```

This `checkSex` function only requires the path containing the raw data in PennCNV format. The function assumes that the LRR variable is in the 4th column. A different column can be specified with `LRRCol`. The process can be parallelized with the argument `mc.cores` that specifies the number of cores to be used.

The result of `checkSex` can be plotted with `plot`. Figure 5.14 shows the

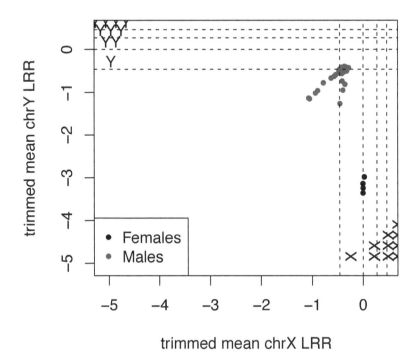

FIGURE 5.14

Cluster of samples using LRR of the chromosome X and Y on individuals belonging to the *brgedata* data example.

LRR for both X and Y chromosomes. In our example, we can observe that males and females are clearly clustered in different groups

```
> plot(sex)
```

The figure shows two additional female samples detected, from the two initially self-reported females. We can summarize the results

```
> sex
Object of class checkSex
---------------------------
Number of processed samples: 126
Number of female classified samples:  4
Number of male classified samples:  122
```

and identify the female samples

```
> sex$par$files[sex$class=="FEMALE"]
[1] "LOY/rawData/FEMALE_2291.txt" "LOY/rawData/FEMALE_2439.txt"
[3] "LOY/rawData/SAMPLE_4.txt"    "LOY/rawData/SAMPLE_88.txt"
```

Females are removed from downstream analyses, selecting the files corresponding to males only

```
> files.males <- sex$par$files[sex$class!="FEMALE"]
```

`files.males` contains the median mLRR-Y for each sample. mLRR-Y can be affected by several artifacts and should be corrected. The overall LRR distribution across samples is expected to be zero, deviations from this systematic error should be removed by normalization of the median mLRR-Y data by the LRR in the autosomes. This can be done with the 5% trimmed-mean of LRR to avoid regions with copy number alterations. In cancer, aneuploidies of Y are common, therefore, the trimmed value of the LRR may be increased up to 25% when analyzing cancer samples. mLRR-Y may be centered at 2/3 `log(1/2) = -0.46` corresponding to 1 copy of chromosome Y (e.g ploidy is equal to 1). However, the calling is improved by removing the median value of the mLRR-Y in all individuals, and centering at 0.

`madloy` computes the median mLRR-Y across samples with respect to a reference region

```
> mLRR <- madloy(files.males)
> mLRR
Object of class MADloy
---------------------------
Number of processed samples: 122
Target region: chrY:6671498-22919969
Reference region(s): Autosomal chromosomes
Offset (median LRR value in msY): -0.56
```

The result is an object of class *MADloy*, in which we observe that LRR data has been summarized in a target and a reference region. The target region corresponds to the mLRR-Y while the reference region, used to normalized LRR in Y, corresponds to autosomal chromosomes. The arguments `target.region` and `ref.region` can be used to change the regions in UCSC format (e.g. "chr21 or "chr21:1000-10000).

The reference human genome can be changed in the argument `hg`, the default is hg18. The package also contains files to retrieve summarized data in X and Y PAR regions, p and q arms, and msY region to help the description of LOY events.

Samples with large variability in LRR are automatically removed from the analysis and the calling returns NA values of summarized mLRR-Y for samples with outlying standard deviation of LRR (> 0.28 from reference chromosome, see `www.illumina.com/content/dam/illumina-marketing/documents/`

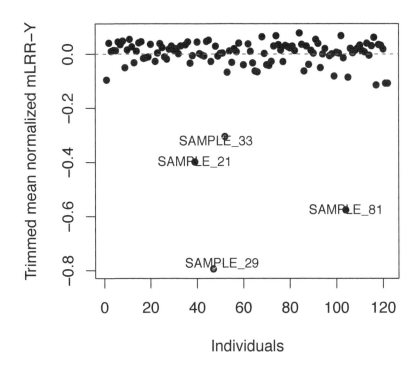

FIGURE 5.15
Plot of MADloy object of males samples from *brgedata*.

`products/appnotes/appnote_cnv_loh.pdf`. This value can be changed by setting `qc.sds`. If missing, samples, having LRR larger than 2 times the LRR standard deviation across all samples, are removed.

Objects of class *MADloy*, `mLRR`, can be visually inspected with the generic `plot` function. Figure 5.15 shows the mean difference between the LRR of the Y chromosome and the LRR across autosomes (reference). It shows numerous samples with likely LOY (those in the -.2, -.8 range of the Y-axis)

```
> plot(mLRR, print.labels=TRUE, threshold=-0.3)
```

We can visually inspect individual samples to confirm LOY status. For instance, figure 5.16 shows a normal individual (SAMPLE_1) where LRR (dots) in the mLRR-Y region (shaded area) does not deviate from the expected value in normal samples (horizontal bottom dashed line). The figure can be obtained by:

```
> plotIndLRR(mLRR, sample="SAMPLE_1")
```

On the other hand, Figure 5.17 shows a case with detectable LOY since the LRR is far below from the reference (bottom dashed line).

```
> plotIndLRR(mLRR, sample="SAMPLE_29")
```

LOY calling in *MADloy* uses both LRR and BAF. It improves other threshold methods such as one implemented *MADloy* based on detecting outliers in LRR distribution or the one proposed by Fosberg and colleagues [36]. These two types of calls as well as the continuous value of LRR in chromosome Y can be obtained by

```
> mLRR.call <- getLOY(mLRR)
> mLRR.call

Object of class LOY
----------------------------
Number of normal samples: 111
Number of LOY: 4
Number of XYY: 0
Number of samples do not pass QC: 7
>
> head(mLRR.call$res)
                MADloy Fosberg     continous
SAMPLE_1.txt    normal     LOY -0.095543539
SAMPLE_10.txt   normal  normal  0.040349783
SAMPLE_100.txt  normal  normal  0.009781045
SAMPLE_101.txt  normal  normal  0.013606487
SAMPLE_102.txt  normal  normal  0.012527824
SAMPLE_103.txt  normal  normal  0.044729874
```

The number of LOY samples provided by Fosberg's method can be obtained by

```
> table(mLRR.call$res$Fosberg)

normal    LOY
   101     14
```

The optimal calling proposed by *MADloy* is performed by

```
> loy <- checkBdev(mLRR.call)
> loy
Object of class MADloyBdev
----------------------------
class
discordant       LOY    normal      other
         4         4       100          7
```

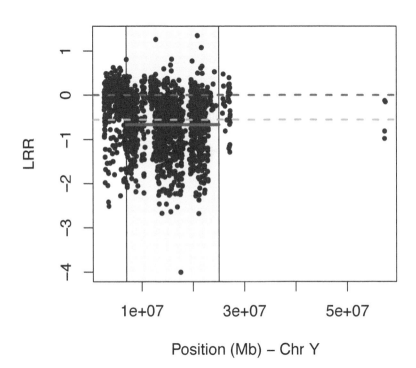

FIGURE 5.16
LRR (dots) in the mLRR-Y region (shaded) of SAMPLE_1 (normal sample)
from *brgedata* data example. The solid horizontal line represents the median
LRR values in the mLRR-Y region while the horizontal bottom dashed line
is the expected value in a normal sample.

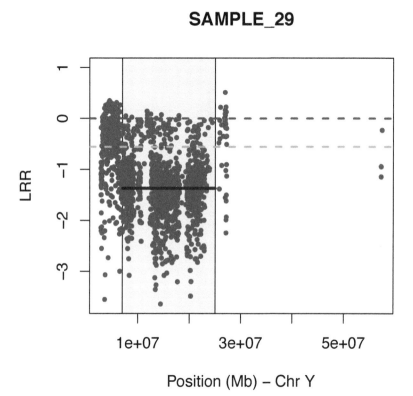

FIGURE 5.17
LRR in the mLRR-Y region (shaded) of SAMPLE_29 from the *brgedata* data
example.

The function indicates that there are 4 individuals having LOY. There are other 4 samples that cannot be considered as LOY since the b-Deviation is discordant with the expected value, and there are other 7 samples having different types of rearrangements. These individuals can be identified by

```
> loyDat <- loy$data
> sel <- which(loyDat$class == "other")
> rownames(loyDat)[sel]
[1] "SAMPLE_111" "SAMPLE_114" "SAMPLE_13"  "SAMPLE_47"  "SAMPLE_67"
[6] "SAMPLE_89"  "SAMPLE_98"
```

Figure 5.18 shows why, although SAMPLE_98 has a decreased LRR in chromosome Y, it cannot be considered a LOY event since it is due to data contamination as the B-deviation is indicating.

```
> plotIndSNPX(mLRR, sample="SAMPLE_98")
```

It is important to notice that functions checkSex, madloy and checkSex have an argument called mc.cores that can be used to paralellize the analyses.

A plot on loy shows individuals with positive LOY, see Figure 5.19. The figure can be improved printing labels with the aid of the *wordcloud* package

```
> plot(mLRR.call, ylim=c(-1.5, 0.5), print.labels=TRUE)
```

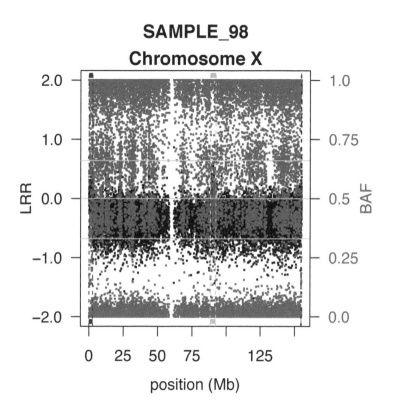

FIGURE 5.18
LRR and BAF of a sample being tagged as 'other' by MADloy calling

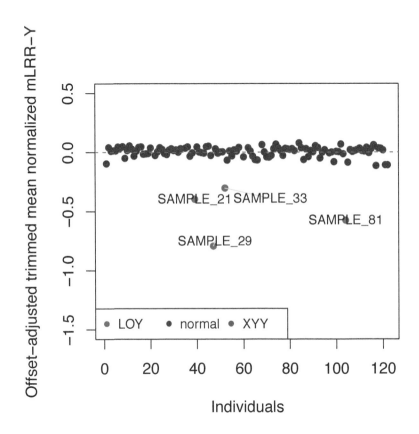

FIGURE 5.19
LOY calling on the *brgedata* data example.

5.5   Polymorphic inversions

Inversion polymorphisms are part of the genome's structural variation repertoire. They were first described by Alfred Sturtevant in 1921 [148] and have
been intensively researched in numerous species, in particular in *Drosophila*.
Inversions have been related to the emergence of chromosome Y in mammals
with important roles in adaptation and evolution.

Inversions polymorphisms are specific genomic segments that run in the
opposite sense of a reference or a wild-type chromosome. Individuals can be
homozygous for the inversion, heterozygous of wild-type homozygous. Large
inversions were firstly described by the microscope and soon one of their main
properties was established: suppression of recombination for heterozygous individuals. The suppression of recombination is thought to contribute to chromosomal diversity, important for an individual's adaptation to changing environments [71]. Recent genomic data, either microarray or sequence based, has
enabled the characterization of submicroscopic inversions in large population
samples [16, 17, 86, 133]. In humans, microscopic inversions are rare, while
submicroscopic are thought to be from hundreds to thousands, many of them
frequent in the general population. While high throughput DNA sequencing
has been important in describing new inversions in few individuals, SNP microarray data offers the possibility to study their population effects. Genotyping inversions from SNP data is challenging. However, there are methods that
can be used to identify and genotype inversions that are old, non-recurrent
and large enough to accumulate specific genomic variability. Here, we show
how inversions can be detected and genotyped from SNP data in large cohorts
and how to perform association tests with phenotypes.

One of the signals that inversions leave on SNP data is the singular changes
in the linkage disequilibrium (LD)across their breakpoints. The genome-wide
search for those LD patterns is used in the Bioconductor package *inveRsion*.
Here, we illustrate how *inveRsion*'s algorithm can be used to detect genomic
regions where inversion may be present. We then explain how to genotype
inversions with *invClust*, for inversions whose breakpoints have been experimentally validated in few individuals. This algorithm is a clustering method for
the multidimensional scaling of the SNPs within the breakpoints. It assumes
that the inversion sustains two divergent haplotype groups, given by the suppression of recombination in the region.*invClustinvClust* is population-based
and, therefore, requires population samples where the clustering is reliable. We
finally describe a third method, *scoreInvHap* to genotype individuals, based
on reference haplotypes for each inversion state. This method allows the quick
genotyping on individuals on specific inversions and it is particularly useful
for multicentric association studies with genomic data obtained in different
platforms.

5.5.1 Inversion detection

We illustrate the use of *inveRsion* to detect a robust inversion signal in humans at 17q21. This inversion has been shown to be under selective pressures since its appearance in Africa, where the ancestral allele almost disappeared. The ancestral allele is, however, the most common in Europeans and has been associated with reproductive fitness in Icelanders [144].

We can download the genotype data from the 1000 Genomes Project, available from the public ftp repository of the genome browser `http://hgdownload.cse.ucsc.edu/gbdb/hg19/1000Genomes/`. The genotype data of the entire chromosome 17 from all the population ancestries in the 1000 Genomes can be obtained in a single file. The data is in a variant call format (vcf) file that can be read into R using the Bioconductor package *variantAnnotation*. We are interested only in the genotypes of the SNPs within the inversion's region, that can be selected using the genomic range defined by the `GRranges` function of *GenomicRanges*. We thus load the SNP genotypes within the inversion with the `readVcf` and parameter `param`

```
> library(VariantAnnotation)
> fl <-
+   "ALL.chr17.phase3_shapeit2_mvncall_integrated_v5a.20130502.genotypes.vcf.gz"
>
> #generate tabix file if needed
> #idx <- indexTabix(fl, "vcf")
> #tab <- TabixFile(fl, idx)
>
> rng <- GRanges(seqnames="17", ranges=IRanges(start=43e6, end=47e6))
> vcf <- readVcf(fl, "hg19",param=rng)
>
```

The SNP genotypes are accessed with the function `geno` and the field `GT`. These are phased genotypes, for which we reconstruct the haplotype of each individual in the European CEU population, setting the corresponding SNP coordinates as column names. We store the haplotypes in the file `hap.txt` that will be used by *inveRsion*,

```
> #extract genotypes
> genos <- geno(vcf)$GT
>
> popinfo <- read.delim("20130606_g1k.ped", as.is=TRUE)
> selCEU <- popinfo$Individual.ID[popinfo$Population=="CEU"]
>
> #select CEUs
> genosCEU <-genos[, colnames(genos)%in%selCEU]
> selout <- rowMeans(genosCEU == "0|0") != 1
> genosCEU <- genosCEU[selout,]
>
> #construct haplotype matrix
> hap1 <- matrix(0, ncol=ncol(genosCEU), nrow=nrow(genosCEU))
> hap1[genosCEU == "1|0"]<-1
> hap1[genosCEU == "1|1"]<-1
>
```

```
> hap2<-matrix(0, ncol=ncol(genosCEU), nrow=nrow(genosCEU))
> hap2[genosCEU == "0|1"]<-1
> hap2[genosCEU == "1|1"]<-1
>
> hap<-lapply(1:ncol(hap1), function(x) rbind(hap1[,x],hap2[,x]))
> hap<-do.call(rbind,hap)
>
> colnames(hap)<-start(rowRanges(vcf))[selout]
>
> #write matrix formatted for inveRsion
> write.table(hap, file="hap.txt",
+             col=TRUE, row=FALSE, quote=FALSE, sep="\t")
>
```

The main source data for the *inveRsion* algorithm is a data table with two rows per individual that represent each of the subject's haplotypes, in the region where we are searching for an inversion. The algorithm fits probabilistic models that test the likelihood that, between two candidate breakpoints, there are some chromosomes that are inverted. The likelihood is written in terms of the linkage disequilibrium differences between the case of zero inverted chromosomes in the data and the optimal case of some chromosomes being inverted. A window of a given size between the breakpoints is defined and then moved across the interrogated genomic segment. Windows with evidence of inversions are given the Bayesian information criterion (BIC) score, derived from the probabilistic model. Positive BICs are positive signals for the inversion.

The segment between two breakpoints is interrogated with four blocks of N SNPs, two flanking blocks at both sides of each breakpoint. The model measures the linkage between the blocks in the forwards and inverted models For efficiency, the blocks between each candidate breakpoint are encoded as decimal numbers in the `codeHaplo` function. As candidate breakpoints are anywhere between any two contiguous SNPs, there are as many candidate breakpoints as SNPs in the data set. One breakpoint can be tested up to two times, one with an inversion of window length to its left and another with an inversion to its right-hand side.

Using the 1000 Genomes data for the 17q21 region written in the `hap.txt`, we code the two N SNP blocks at each side of two contiguous SNPs using `codeHaplo`, with parameters `blockSize=N` (default 5), `minAllele` (minimum allele frequency of SNPs in the blocks) and `saveRes` (whether to save result).

```
> library(inveRsion)
>
> hapCode<-codeHaplo(file="hap.txt",
+                    blockSize=5, minAllele=0.1,
+                    saveRes=FALSE)
```

Single breakpoints are thus encoded by their flanking SNP blocks. Inversion models are tested for each pair of breakpoints a `window` distance given in megabases (i.e. 0.5), using the function `scanInv` function. The parameter `geno` is given when non-phased SNP genotypes are used. While the function can handle local phasing of the genotypes, current algorithms for phasing, such as SHAPEIT, are efficient and can be used on the original genotypes.

```
> window <- 0.5
> scanRes <- scanInv(hapCode,window=window,
+                    saveRes=FALSE, geno=FALSE)
>
> invList <- listInv(scanRes, hapCode=hapCode,
+                     geno=FALSE,all=FALSE,thBic=0)
>
> invList
```

The `listInv` function summarizes the significant models (with $BIC > thBic$) that are gathered into overlapping signals. In the region interrogated, we can see a clear signal obtained in the inverted region, where an estimation of the inversion frequency is given together with the number of models with positive BIC, as evidence for the existence of a frequent inversion in the population sample. The result can be better visualized with the plotting of the scanning results.

```
> plot(scanRes)
```

5.5.2 Inversion calling

When the inversion breakpoints are known and have been experimentally validated, old non-recurrent inversions can be reliably genotyped in individuals. Population genetics and association studies can then be carried out. We illustrate how to call the inversion genotypes of the largest common inversion found in humans: inversion 8p23. We choose the SNP genotypes of the European individuals, CEU, as obtained in the HapMap project, with microarray data of over one million SNPs across the genome. The SNP coverage is less dense that for the 1000 Genomes but can be used to reliably call the 8p23 inversion genotypes. The data is available in PLINK format from the NCBI ftp repository `ftp://ftp.ncbi.nlm.nih.gov/hapmap/genotypes/`.

Data can be loaded in R with **read.plink** of *snpStats*. We select the CEU individuals in the data and format the SNP annotation information to be passed to the *invClust* algorithm

```
> library(snpStats)
> fl <- "hapmap3_r2_b36_fwd.consensus.qc.poly"
> genosHapMap <- read.plink(fl)
>
> #select CEU
> popinfo <- read.delim(file="relationships_w_pops_121708.txt",
+                       as.is=TRUE)
> selpops <- popinfo$population%in%"CEU"
> ceusIDs <- rownames(genosHapMap$genotypes)%in%popinfo$IID[selpops]
```

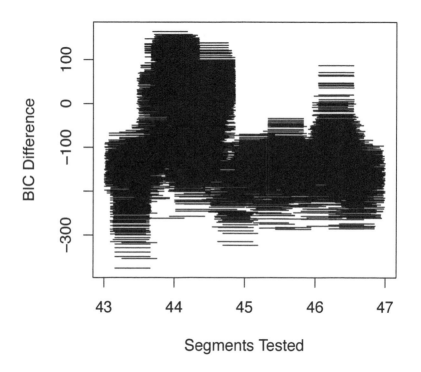

FIGURE 5.20
Scanning of the 17q21 region for signal of an inversion in the CEU population
of the 1000 genomes.

```
>
> geno <- genosHapMap$genotypes[ceusIDs,]
> annot <- genosHapMap$map[c("chromosome", "snp.name", "position")]
```

invClust takes the SNP genotypes, and their annotation, in the region within the inverted segment and performs a multidimensional scaling. It has been shown that inversions support divergent haplotype groups when no recurrence has occurred. When the haplotypes within one inversion have accumulated enough variability then individuals are clearly classified into haplotype-genotypes, each corresponding to an inversion genotype. In the first components of a multidimensional scaling of the SNPs within ancestral nonrecurrent inversions, we thus expect individuals to cluster into three equidistant groups; the groups at the edges being standard and inversion homozygotes and the group in the middle being inversion heterozygotes. This neat mapping is observed in the CEU individuals of the HapMap for inversion 8p23.

invClust requires the region of interest (ROI) between which inversion polymorphism takes place. Then the algorithm utilizes up to two multidimensional components, specified in dim, to estimate the clustering of the individuals. The clustering is a probabilistic model in which Hardy–Weinberg equilibrium is assumed between the inversion alleles. This reasonable condition for polymorphic inversions improves classification, particularly, in cases where there is overlap in the genotype groups

```
> library(invClust)
>
> roi <- data.frame(chr=8, LBP=7897515, RBP=11787032, reg="8p23")
>
> invCall <- invClust(roi=roi, wh = 1, geno=geno, annot=annot, dim=2)
```

As a result, the algorithm produces an invClust object, reporting the inversion frequency and the amount of variability explained by the components computed in the multidimensional scaling step. The inversion genotypes for each individual are stored in the genotypes field of the invClust object

```
> invCall
Inversion genotype clustering
-object of class invClust-
  fields: $EMestimate: mixture model parameters
          $datin: fitted data
  subjects: 165
  groups fitted: 1
  overall inversion allele frequency: 0.381783
  variance explained by  5   MDS componet(s): 0.864447
> head(invCall["genotypes"])
        NI/NI NI/I I/I
NA06989     0    0   1
NA11891     0    1   0
NA11843     0    1   0
NA12341     1    0   0
NA12739     1    0   0
NA10850     1    0   0
```

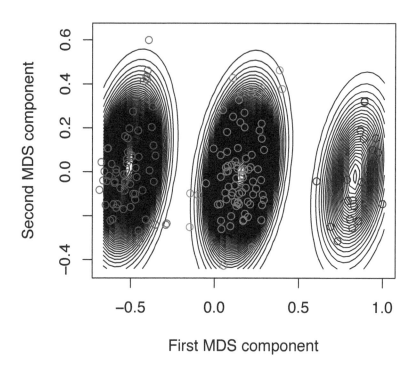

FIGURE 5.21
Call of inv-8p23 in the CEU population of HapMap using invClust.

The clear clustering into the three inversion-genotype groups of inversion 8p23 is observed for the CEU individuals

```
> plot(invCall)
```

This genotype calling is population-based and is sensitive to ancestral differences in the population sample. If substantial variability has accumulated between common inversion states over different ancestries then the clustering has to account for these differences. In general practice, if the inverted allele is known for different individuals then the clustering stratifying by ancestry is sufficient to call the inversion genotypes. This approach has been used to study the global distribution of 8p23 and detect selection signals acting upon it.

5.5.3 Inversion association

When a set of individuals are known with experimentally validated inversions for each genotype and their SNP haplotypes can be reconstructed, then, it is possible to perform an inversion call that is individually based. *inveRsion* and *invClust* are algorithms based in population statistics. Sufficiently large groups are needed to detect the inversion and perform the genotype calling. With a reference group of individuals, for each inversion genotype, it is possible to map a new single individual to the groups using a score implemented in *scoreInvHap*. The score is flexible to include only the SNPs available for the individual, which may be few in comparison with the reference. This allows the genotype calling on individuals with different SNP coverages, simplifying the study of inversions in multicentric association studies, where subjects have been genotyped in different platforms.

We illustrate the association between 8p23 inversion and obesity, as the SNP rs17150703 in *MSRA* within 8p23 has been associated with BMI in adults [135]. We use the obesity data of Section 4.5.4. We first recover the PCA performed in the quality control of the data to adjust the association for population stratification. In the selected subjects who passed quality control, we call the inversion genotypes.

```
> #load genotypes
> library(snpStats)
> genos.ob <- read.plink("obesity")
>
> #recover PCA analysis
> load(file="pca.RData")
>
> eig <- data.frame(c1=pca$eigenvect[,1], c2=pca$eigenvect[,1])
> rownames(eig)<-as.character(pca$sample.id)
>
> #select individuals
> selids<-rownames(genos.ob$genotypes)%in%as.character(pca$sample.id)
>
> genos.ob$genotypes<-genos.ob$genotypes[selids,]
>
> #confirm subject selection
> identical(rownames(genos.ob$genotypes),as.character(pca$sample.id))
[1] TRUE
```

We then use **scoreInvHap** to perform the inversion call of 8p23, with respect to experimental references (**invClust** could also be used). The function is implemented with references for 20 human inversions, including 8p23 (named **inv8_001**), and with instructions on how to create the reference for new inversions in a Bioconductor package [128]. The mapping of individuals into the reference genotypes is controlled by selecting SNPs in high linkage disequilibriumwith the inversion (**SNPsR2**) and by their frequencies in the reference groups [15].

```
> library(scoreInvHap)
>
> callinv <- scoreInvHap(genos.ob,
+                        SNPsR2 = SNPsR2$inv8_001,
+                        hetRefs = hetRefs$inv8_001,
+                        Refs = Refs$inv8_001)
>
> inv8p23 <- classification(callinv)
>
> head(inv8p23)
 100 1001 1004 1005 1006 1008
  II   NI   NI   II   NI   II
Levels: II NI NN
```

Having called the inversion genotypes in the obesity cohort, we test the association between the inversion and obesity. We adjust by the first two genome-wide PCA components, country and smoke status. We use *SNPassoc* to test the association (see Chapter 4)

```
> library(SNPassoc)
>
> #load obesity phenotypes and merge data with 8p23 inversion call
> ob.pheno <- read.delim(file="obesity.txt")
> rownames(ob.pheno) <- ob.pheno$id
>
> idspheno <- intersect(names(inv8p23), rownames(ob.pheno))
>
> datainv <- data.frame(inv=inv8p23[idspheno], eig[idspheno,],
+                       ob.pheno[idspheno,])
>
> #set up data for SNPassoc
> ob.inv <- setupSNP(data=datainv, colSNPs=1, sep="")
>
> #association test
> association(obese ~ inv + c1+ c2+ country + smoke, ob.inv)
```

SNP: inv adjusted by: c1 c2 country smoke									
	0	%	1	%	OR	lower	upper	p-value	AIC
Codominant									
I/I	565	32.5	128	32.2	1.00			0.10193	2040
N/I	819	47.2	170	42.7	0.93	0.72	1.19		
N/N	352	20.3	100	25.1	1.26	0.93	1.69		
Dominant									
I/I	565	32.5	128	32.2	1.00			0.84377	2043
N/I-N/N	1171	67.5	270	67.8	1.02	0.81	1.29		
Recessive									
I/I-N/I	1384	79.7	298	74.9	1.00			0.04019	2039
N/N	352	20.3	100	25.1	1.32	1.02	1.70		
Overdominant									
I/I-N/N	917	52.8	228	57.3	1.00			0.12881	2041
N/I	819	47.2	170	42.7	0.84	0.68	1.05		
log-Additive									
0,1,2	1736	81.3	398	18.7	1.11	0.95	1.29	0.19320	2041

We, therefore, find a significant association between obesity and the recessive model for the inversion. Consistent with previously reported findings, the results further indicate the involvement of the inversion in the reported associations with BMI [16]. This observation motivates further analyses to investigate whether the association between specific SNPs and obesity is mediated by the inversion. Inversion polymorphisms are an important source of genomic variation. The ability to investigate their effects in adaptation, evolution and phenotypic differences between individuals is determined by the possibility to infer them in large population samples. We have shown how this can be achieved for two human inversions, using genomic SNP data.

6

Addressing batch effects

CONTENTS

6.1   Chapter overview

Omic data are the product of high throughput technologies that are affected by laboratory conditions, reagent lots and personnel. Biological systems are highly reactive on small changes in the surrounding physical an chemical conditions. Therefore, *omic* data depends on the date and place of processing as surrogate variables of uncontrolled conditions. Samples that are processed in the same laboratory and the same place will reflect such batch effect in their data. All types of *omic* data are affected by batch effects. While the effects can be mitigated by a suitable study design, they cannot be completely removed [79]. Clearly, if all cases are processed together, their differences with controls cannot be teased apart from the batch effect. In addition, while statistical estimates can be adjusted by laboratory and date, there may be measurements in some genes that are more reactive than others and will be more subjected to confounding than others.

In this chapter, we analyze already collected data and readers interested in collecting new data should refer to authors discussing the study design of specific *omic* studies. In this chapter, we illustrate how to detect and correct the batch effect, taking as an example transcriptomic data. We discuss how to detect unwanted variation in *omic* data from high throughput experiments using surrogate variable analysis (SVA) [78]. When batch effect variables have been reported then corrected datasets can be obtained using the ComBat algorithm [66].

6.2   SVA

We show how to detect the batch effect in the transcriptomic study of Alzheimer's disease available in GEO with accession number GSE63061, introduced in Chapter 2. This is a study that aims to detect transcriptomic differences, using microarray data, among Alzheimer's disease cases (AD), mild cognitive impairment patients (MCI) and controls (CTL).

As previously described, we download data from GEO using the `getGEO` function from the *GEOquerry* package:

```
> library(GEOquery)
> gsm.expr <- getGEO("GSE63061", destdir = ".")
> gsm.expr <- gsm.expr[[1]]
> show(gsm.expr)



ExpressionSet (storageMode: lockedEnvironment)
assayData: 32049 features, 388 samples
  element names: exprs
protocolData: none
phenoData
  sampleNames: GSM1539409 GSM1539410 ... GSM1539796 (388 total)
  varLabels: title geo_accession ... data_row_count (34 total)
  varMetadata: labelDescription
featureData
  featureNames: ILMN_1343291 ILMN_1343295 ... ILMN_3311190 (32049
    total)
  fvarLabels: ID Species ... GB_ACC (30 total)
  fvarMetadata: Column Description labelDescription
experimentData: use 'experimentData(object)'
Annotation: GPL10558
```

We have downloaded an *ExpressionSet* object that contains both gene expression and phenotypic data. Gene expression can be obtained with the `exprs` function as described in Chapter 3

```
> expr <- exprs(gsm.expr)
> dim(expr)
[1] 32049    388
```

while phenotype data is obtained with **pData**. For this study the case/control status of the individuals is encoded in the variable `characteristics_ch1` of the **pData** object.

```
> #get phenotype data
> pheno <- pData(phenoData(gsm.expr))
> status <- pheno$characteristics_ch1
> status <- gsub("status: ","", as.character(status))
> table(status)
status
```

AD borderline MCI		CTL	CTL to AD	MCI
139	3	134	1	109

MCI to CTL	OTHER			
1	1			

We save the subject status into the factor variable **fstatus**. For simplicity, we relabel the levels into case/control, including MCI and OTHER in cases.

```
> fstatus <- factor(status)
> levels(fstatus) <- c("case","case", "cont", "cont", "case", "case","case")
> table(fstatus)
fstatus
case cont
 253  135
```

SVA is a Bioconductor's package to detect surrogate variables in different types of *omic* data. *SVA* extracts surrogate variables comparing the transcriptomic data as modeled by the covariates only (null model **mod0**) with the data modeled by the covariates and the variables of interest (full model **mod**). In our data, we do not have covariates and therefore the null model is given only by the intercept

```
> library(sva)
> mod0 <- model.matrix( ~ 1, data = fstatus)
> mod <- model.matrix( ~ fstatus)
```

The function **num.sv** estimates the number of surrogate variables latent in the model.

```
> n.sv <- num.sv(expr, mod, method="leek")
> n.sv
[1] 0
```

We observe that this data does not seem to contain any significant surrogate variable. To confirm this, we compute the first two surrogate variables with the function **sva** and parameter **n.sv=2**

```
> svobj <- sva(expr, mod, mod0, n.sv=2)
Number of significant surrogate variables is:  2
Iteration (out of 5 ):1  2  3  4  5
> names(svobj)
[1] "sv"        "pprob.gam" "pprob.b"   "n.sv"
```

the surrogate variables are stored in the variable **sv** which can be plotted and compared to the subject status

```
> col <- fstatus
> levels(col) <- c("black","grey")
> plot(svobj$sv[,1:2],col=as.character(col), xlab="sva 1",
+      ylab="sva 2", pch=16)
```

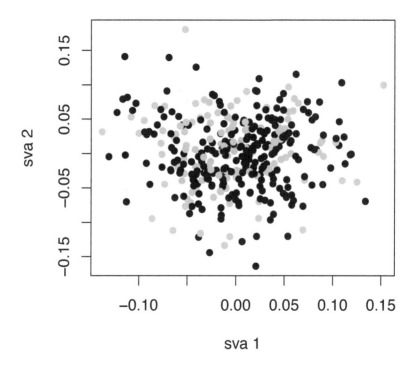

FIGURE 6.1
SVA components of GSE63061. Cases are black dots and controls are grey.

The plot confirms that there are no latent variables associated with the case-control status. We compare this result with the one obtained from unprocessed data from the same study. Unnormalized expression data can be downloaded from the GEO's web-page of GSE63061. We load the data into the **data.frame exprNN**, using the R function **read.table**, and apply a log_2 transformation

```
> exprNN <- read.table("GSE63061_non-normalized.txt",
+                       header=TRUE, check.names=FALSE, as.is=TRUE)
> probesID <- exprNN[,1]
> selprobes <- probesID%in%rownames(expr)
>
> exprNN <- exprNN[selprobes,as.character(pheno$title)]
> exprNN <- log2(exprNN)
```

Expression data for **exprNN** is ordered across subjects, with respect to the processed data **expr**, using the subject identifiers in the phenotype data under the variable **pheno$title**. For this case, we see there are two surrogate variables in the data

```
> n.sv <- num.sv(exprNN, mod, method="leek")
> n.sv
[1] 2
```

We thus compute the number of surrogate variables identified

```
> exprNNmat <- as.matrix(exprNN)
> svobjN <- sva(exprNNmat, mod, mod0, n.sv=n.sv)
Number of significant surrogate variables is:  2
Iteration (out of 5 ):1  2  3  4  5
```

and plot them against each other with the underlying case-control status

```
> plot(svobjN$sv[,1:2],col=as.character(col), xlab="sva 1",
+      ylab="sva 2", pch=16)
```

In this case, we see that the surrogate variables have more structure. In particular, a cluster appears in the second SVA variable, which shows an association trend with Alzheimer's disease diagnosis

```
> ADstatus <- factor(status)
> levels(ADstatus)
[1] "AD"             "borderline MCI" "CTL"            "CTL to AD"
[5] "MCI"            "MCI to CTL"     "OTHER"
> levels(ADstatus) <- c("1","0", "0", "0", "0", "0","1")
>
> summary(glm(ADstatus ~ svobjN$sv, family="binomial" ))

Call:
glm(formula = ADstatus ~ svobjN$sv, family = "binomial")

Deviance Residuals:
```

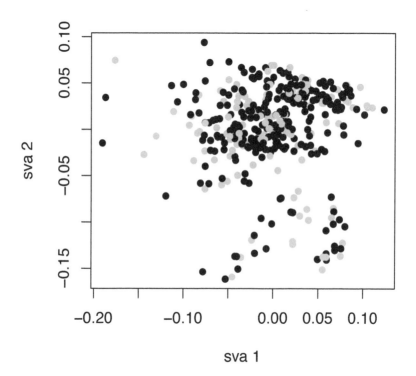

FIGURE 6.2
SVA components of unnormalized data for GSE63061. Cases are black dots
and controls are grey.

```
   Min      1Q   Median      3Q      Max
-1.7413  -1.3449   0.8904   0.9665   1.0761


Coefficients:
            Estimate Std. Error z value Pr(>|z|)
(Intercept)   0.5795     0.1066   5.437 5.42e-08 ***
svobjN$sv1   -1.5087     2.1067  -0.716   0.4739
svobjN$sv2   -4.1877     2.2050  -1.899   0.0575 .
---
Signif. codes:  0 '***' 0.001 '**' 0.01 '*' 0.05 '.' 0.1 ' ' 1

(Dispersion parameter for binomial family taken to be 1)

    Null deviance: 507.42  on 387  degrees of freedom
Residual deviance: 503.11  on 385  degrees of freedom
AIC: 509.11

Number of Fisher Scoring iterations: 4
```

We thus see that unprocessed data show a batch effect that can affect differences between the different diagnosis statuses of subjects. Not accounting for this effect can thus lead to false positive results. Note however that the surrogate variables that correspond to the processed data, while they do not associate with the case-control status defined in `fstatus`, they do associate with the MCI status of the original phenotype `status`.

```
> summary(glm(svobj$sv[,1]  ~ status))

Call:
glm(formula = svobj$sv[, 1] ~ status)

Deviance Residuals:
      Min        1Q    Median        3Q       Max
-0.132372  -0.033532  -0.000651   0.034718   0.157782

Coefficients:
                        Estimate Std. Error t value Pr(>|t|)
(Intercept)            -0.001930   0.004265  -0.452   0.6512
statusborderline MCI   -0.074370   0.029342  -2.535   0.0117 *
statusCTL              -0.003562   0.006088  -0.585   0.5588
statusCTL to AD        -0.013579   0.050463  -0.269   0.7880
statusMCI               0.013240   0.006433   2.058   0.0403 *
statusMCI to CTL       -0.003673   0.050463  -0.073   0.9420
statusOTHER             0.023247   0.050463   0.461   0.6453
---
Signif. codes:  0 '***' 0.001 '**' 0.01 '*' 0.05 '.' 0.1 ' ' 1

(Dispersion parameter for gaussian family taken to be 0.002528358)

    Null deviance: 1.0000  on 387  degrees of freedom
Residual deviance: 0.9633  on 381  degrees of freedom
AIC: -1210.3

Number of Fisher Scoring iterations: 2
```

Therefore, analysis of the processed data that aims to find differences between subjects still needs to adjust for surrogate variables to avoid false positives.

6.3   ComBat

`ComBat` function implements an algorithm that allows the creation of a new expression dataset in which the batch effect has been removed. Opposite to SVA, ComBat requires the batch effects to be known and are typically encoded in the date of processing or the laboratory label. Here, we analyze another unprocessed dataset of human cortical gene expression for which the batch effect variable has been reported. The dataset corresponds to the GEO accession number GSE8919. Unprocessed data is accessible at the study website `http://labs.med.miami.edu/myers/LFuN/LFuN.html`. Downloaded data can be uploaded in an R session with `read.table`, where the first column corresponds to the probe IDS and expression data is the log_2 transformed

```
> expr <- read.table(file="expression_data_Brain.txt", header=TRUE,
+                    as.is=TRUE )
>
> probesIDS<-expr[,1]
> expr <- as.matrix(log2(expr[,-1]))
>
> expr[1:5,1:5]
     wgacon.219 wgacon.221 wgacon.234 wgacon.274 wgacon.277
[1,]        NaN   4.852998        NaN        NaN        NaN
[2,]  13.319559  12.834965  12.181308  12.471421  12.463601
[3,]   7.033423   6.416164   7.733354   6.854245   6.963474
[4,]        NaN   5.491853        NaN        NaN        NaN
[5,]  10.856192  10.338179  10.329236   9.978853   9.939579
```

`ComBat` can deal with missing variables when not all subjects in a batch are missing for a given probe. For simplicity, we remove probes with any missing values across subjects

```
> selna<-!is.na(rowSums(expr))
> expr<-expr[selna,]
```

Covariates are accessible in a separate text file, in which a batch variable is included with name `institute_sample_source`

```
> cov <- read.table(file="covariate_data_Brain.txt", header=TRUE,
+                    as.is=TRUE )
> names(cov)
 [1] "Group"                  "Member"
 [3] "gender"                 "age_at_death"
 [5] "pmi"                    "transcripts_detected_rate.24354."
```

```
 [7] "trans_det_rate.14078."              "exp_hyb_date"
 [9] "institute_sample_source"            "brain_region"
> table(cov$institute_sample_source)

 A  B  C  D  E  F  G  H  I  J  K  L  M  N  O  P  Q  R
21  4 19  3  9 12 34 16  9  1  3  4 11 11  6 14 14  2
```

A matrix for the null model that includes other covariates is defined and included in the ComBat that produces an expression dataset where the batch variation has been removed

```
> modcombat <- model.matrix(~1, data=cov)
>
> expr_corrected <- ComBat(dat=expr, batch=cov$institute_sample_source,
+                          mod=modcombat, par.prior=TRUE, prior.plots=FALSE)
Standardizing Data across genes
>
> expr_corrected[1:5,1:5]
      wgacon.219 wgacon.221 wgacon.234 wgacon.274 wgacon.277
[1,]   13.357321  12.872727  12.219070  12.509183  12.501363
[2,]   10.885134  10.367121  10.358178  10.007796   9.968521
[3,]    7.497673   7.385780   8.315536   8.600793   8.440012
[4,]   10.680490   9.454634   9.618876   9.074535   9.716808
[5,]    7.325802   6.353513  10.251341   7.754545   6.159152
```

We now compare the surrogate variables associated with the corrected and the uncorrected expression datasets. We use the brain region as the variable of interest to define the full and the null models

```
> fbrain <- as.factor(cov$brain_region)
> table(fbrain)
fbrain
 frontal parietal temporal  unknown
      41        3      136       13
>
> mod0 <- model.matrix( ~ 1, data = fbrain)
> mod <- model.matrix( ~ fbrain)
```

For the uncorrected expression dataset **expr**, we find two surrogate variables

```
> n.sv <- num.sv(expr, mod, method="leek")
> n.sv
[1] 4
> svobj <- sva(expr, mod, mod0, n.sv=n.sv)
Number of significant surrogate variables is:  4
Iteration (out of 5 ):1  2  3  4  5
```

as for the corrected dataset **expr_corrected**

```
> n.sv_corrected <- num.sv(expr_corrected, mod, method="leek")
> n.sv
[1] 4
```

```
> svobj_corrected <- sva(expr_corrected, mod, mod0, n.sv=2)
Number of significant surrogate variables is:  2
Iteration (out of 5 ):1  2  3  4  5
```

We see that the batch effect in `institute_sample_source` is strongly associated with the second sva component of the uncorrected dataset, as observed from an analysis of variance of the component with the batch effect variable

```
> batch <-factor(cov$institute_sample_source)
> summary(aov(lm(svobj$sv[,2] ~ batch)))
             Df Sum Sq  Mean Sq F value Pr(>F)
batch        17 0.5305 0.031204   11.63 <2e-16 ***
Residuals   175 0.4695 0.002683
---
Signif. codes:  0 '***' 0.001 '**' 0.01 '*' 0.05 '.' 0.1 ' ' 1
```

For the corrected dataset, while we observe a strong reduction of the association there is still a large amount of unobserved variation that is associated with the batch effect

```
> summary(aov(lm(svobj_corrected$sv[,2] ~ batch)))
             Df Sum Sq  Mean Sq F value  Pr(>F)
batch        17 0.3382 0.019893    5.26 2.7e-09 ***
Residuals   175 0.6618 0.003782
---
Signif. codes:  0 '***' 0.001 '**' 0.01 '*' 0.05 '.' 0.1 ' ' 1
```

Therefore, despite correction for a known batch, surrogate variable detection is still needed to account for unwanted variation.

```
> par(mfrow=c(2,2))
> boxplot(svobj$sv[,1] ~ batch, ylab="sva 1", ylim=c(-0.2,0.2),
+         xlab="batch", main="Uncorrected")
> lines(c(0,20),c(0,0), lty=2)
>
> boxplot(svobj_corrected$sv[,1] ~ batch, ylab="sva 1", ylim=c(-0.2,0.2),
+         xlab="batch", main="Corrected")
> lines(c(0,20),c(0,0), lty=2)
>
> boxplot(svobj$sv[,2] ~ batch, ylab="sva 2", ylim=c(-0.2,0.2),
+         xlab="batch",main="Uncorrected")
> lines(c(0,20),c(0,0), lty=2)
>
> boxplot(svobj_corrected$sv[,2] ~ batch, ylab="sva 2", ylim=c(-0.2,0.2),
+         xlab="batch", main="Corrected")
> lines(c(0,20),c(0,0), lty=2)
```

Batch effects are an important source of *omic* variation and constitute an important source of confounding in association tests. They need to be tackled at all stages of the study, including design, preprocessing and analysis.

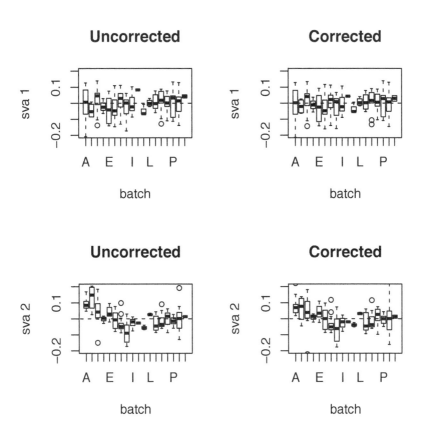

FIGURE 6.3
SVA components of corrected and uncorrected for batch effects the expression data of GSE8919. SVAs of corrected data are more uniform across batches despite differences between them being still present.

7

Transcriptomic studies

CONTENTS

7.1 Chapter overview

The aim of transcriptome-wide association studies is to identify genes whose transcription volume associates with phenotypic differences among individuals. Transcription is measured from tRNA content in biological samples. The transcriptomic data are therefore high-dimensional measurements of the transcriptome under given conditions, such as tissue, age, health status or intervention stage. The transcriptome can be assessed using microarray data or RNA-sequencing (RNA-seq). Microarray data depends on the probes used to detect the amount of transcription of a gene. While microarrays probe gene exons, initial data contains one probe per gene. The amount or intensity of hybridization to a gene's probe measures the transcription output of the gene. If more than one probe is available for a gene, a winsorized mean across probes is taken as the transcription measure of the gene. By contrast, RNA-seq is independent of the probes as it massively sequences short tRNA reads that are then mapped to known gene transcripts. The quantity of reads mapped to a gene is a measure of the transcription of the gene. In this chapter, we will analyze publicly available transcriptomic data based on microarray and RNA-seq experiments. We will discuss how to normalize both types of data and the methods to infer transcriptome-wide association studies.

7.2 Microarray data

Microarrays have been extensively used to detect transcriptomic differences of genes between conditions. Different types of data can be produced from different technologies, each with their own protocols for quality control and preprocessing. Our approach is to illustrate common procedures in the analysis of such data, in particular, those available in public repositories. The *limma* package from Bioconductor is one of the main software resources that allows a wide range of analyses, including the pre-processing of specific microarray technologies and association tests. We refer the reader to the documentation of *limma* for the analysis of newly collected data. Here, we illustrate a typical use of *limma* in the analysis of GEO datasets, following our example of the Alzheimer's disease study GSE63061.

7.2.1 Normalization

An important issue in the analysis of transcriptomic microarray data is the fact that each individual hybridizes differently across the transcriptome probes. This results in different distributions of hybridization intensity across individuals and therefore in an additional variability that is not of biological interest.

Normalization of the distributions refers to the method that makes hybridization intensities as comparable as possible. This type of normalization is between arrays or subjects, in contrast to within-subject normalization that is used in studies that contain replicate per individuals and control probes for which no effect is expected. Transcriptome-wide studies do not typically contain replicates or control probes and therefore normalization is performed between subjects.

We analyze the normalization of the GSE63061 study for which nonnormalized data is available in its GEO's webpage. We retrieve the data and upload it in an R session using the base function `read.table`

```
> exprMat<- read.table("GSE63061_non-normalized.txt",
+                      header=TRUE, check.names=FALSE,
+                      as.is=TRUE)
>
> probesID<-exprMat[,1]
> exprMat<-exprMat[,-1]
> subsID<-colnames(exprMat)
> exprMat<-as.matrix(exprMat)
```

We save the probes and the subject identifiers (IDs) in the variables **probesID** and **subsID**. The function **normalizeBetweenArrays** from *limma* performs the between subject normalization using the **quantile** method

```
> library(limma)
>
```

```
> exprNormdat<-normalizeBetweenArrays(exprMat, method="quantile")
>
> exprNormdat[1:5,1:5]
      7196843081_A 7196843081_B 7196843081_C 7196843081_D 7196843081_E
[1,]      88.85335     85.77075     79.56997     86.27036     79.49330
[2,]     104.12784     89.87732     89.50077     89.19369     89.35103
[3,]      83.83454     82.31160     80.41856     83.18454     81.66031
[4,]      82.41186     81.32358     79.46005     83.91753     81.84201
[5,]      80.20232     84.24369     78.93711     84.66881     81.46302
```

producing a new expression dataset with normalized data. The effect of the normalization can be visualized using the **plotDensities**. We thus concatenate the unnormalized and the normalized data, after a log_2 transformation, into a larger expression dataset and create a subject label variable **gr** to distinguish data from each subset.

```
> L2exprMat<-log2(exprMat)
> L2exprNorm<-log2(exprNormdat)
>
> comp<-cbind(L2exprMat,L2exprNorm)
> nsubs<-ncol(L2exprMat)
>
> gr<-factor(rep(c("Not Normalized","Normalized"),each=nsubs))
```

The plot densities for nonnormalized and normalized data can thus be simultaneously displayed

```
> plotDensities(comp, group = gr, col=c("black","grey"), legend="topright")
```

We observe the large variability in the intensity distributions for the nonnormalized data. Normalization recreates the same distribution across subjects.

A main Bioconductor's object to store expression data is **ExpressionSet**, from the *Biobase* package. It includes the expression intensities, and the phenotypic and probe information. We construct an **ExpressionSet** object with the normalized expression data we have just obtained

```
> library(Biobase)
>
> phenoDF <- data.frame(subsID, row.names=subsID)
> phenodat <- AnnotatedDataFrame(data=phenoDF)
>
> featureDF <- data.frame(probesID, row.names=probesID)
> featureData <- AnnotatedDataFrame(data=featureDF)
>
> exprNorm <- ExpressionSet(assayData=L2exprNorm,
+                           phenoData=phenodat,
+                           featureData=featureData)
>
> show(exprNorm)
ExpressionSet (storageMode: lockedEnvironment)
assayData: 47323 features, 388 samples
```

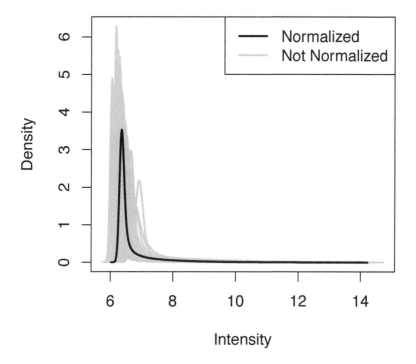

FIGURE 7.1
Distribution of probe intensities for each subject in the expression data set
GSE63061. In grey, the not normalized distribution and in black the normalized ones.

```
   element names: exprs
protocolData: none
phenoData
   sampleNames: 7196843081_A 7196843081_B ... 7348931013_L (388
      total)
   varLabels: subsID
   varMetadata: labelDescription
featureData
   featureNames: ILMN_1762337 ILMN_2055271 ... ILMN_2137536 (47323
      total)
   fvarLabels: probesID
   fvarMetadata: labelDescription
experimentData: use 'experimentData(object)'
Annotation:
```

In `ExpressionSet` the expression data is a matrix while the phenotype data are `AnnotatedDataFrame` which are extended `data.frames` encoding a collection of samples and the values of variables measured in those samples. A meta-data description of the variables can also be included. To access the different types of data the functions `expr`, `pData` and `fData` are used.

We now download the processed data for GSE63061, to be compared with our newly created `exprNorm`. The reported data is accessed using the *GEO-query* package that retrieves the data directly into the R session

```
> library(GEOquery)
> gsm.expr <- getGEO("GSE63061", destdir = ".")
> gsm.expr <- gsm.expr[[1]]
> show(gsm.expr)
```

```
ExpressionSet (storageMode: lockedEnvironment)
assayData: 32049 features, 388 samples
   element names: exprs
protocolData: none
phenoData
   sampleNames: GSM1539409 GSM1539410 ... GSM1539796 (388 total)
   varLabels: title geo_accession ... data_row_count (34 total)
   varMetadata: labelDescription
featureData
   featureNames: ILMN_1343291 ILMN_1343295 ... ILMN_3311190 (32049
      total)
   fvarLabels: ID Species ... GB_ACC (30 total)
   fvarMetadata: Column Description labelDescription
experimentData: use 'experimentData(object)'
Annotation: GPL10558
```

As we can see, this is an `ExpressionSet` object. The probes identifiers for the `gsm.expr` and `exprNorm` are accessed with the `fData` in the variables ID and `probesID`, respectvely

```
> probesIDsGEO<-as.character(fData(gsm.expr)$ID)
> probesID<-as.character(fData(exprNorm)$probesID)
```

Our normalized data `exprNorm` has not been filtered and contain many more probes than the reported `gsm.expr`. For comparison between the datasets, we select common probes

```
> selprobes<-probesID%in%probesIDsGEO
> nselprobes<-sum(selprobes)
>
> exprNorm.sel<-exprNorm[selprobes,]
```

We then want to compare the intensity distributions of the subjects between the two datasets, as we did before using `plotDensities`. We thus concatenate the datasets and create a label variable `gr`

```
> exprGEO<-exprs(gsm.expr)
> expr<-exprs(exprNorm.sel)
>
> comp<-cbind(exprGEO,expr)
>
> nsubs<-ncol(exprGEO)
> gr<-factor(rep(c("Normalized GEO","Normalized"),each=nsubs))
>
> plotDensities(comp, group = gr, col=c("black", "grey"), legend="topright")
```

We see that the distributions in each dataset are normalized and show the same form. However, they are displaced between each other. This systematic difference is typically not of interest as expression levels are relative quantities and no physical units are assigned to them. Given that intensities are relative measurements, intercept estimates in the regression models with the phenotype are not of interest. Displacing all distributions such that their means are zero shows that their shapes are comparable

```
> exprScaledGEO <- scale(exprGEO, scale=FALSE)
> exprScaled <- scale(expr, scale=FALSE)
>
> comp<-cbind(exprScaledGEO,exprScaled)
> plotDensities(comp, group = gr, col=c("black", "grey"), legend="topright")
```

Consequently, both datasets are equivalent and should provide similar results. We, thus reproduce the normalization of this data as available in GEO.

7.2.2 Filter

limma user's manual recommends filtering probes after normalization. Filtering should remove probes with low rates of hybridization compared with the background intensity. Different microarrays and specifications account for the quality of the hybridization in different ways. Here, we show what types of probes were likely filtered out in the GSE63061 as reported in GEO.

We retake the normalized dataset `exprNorm` defined in the previous section to extract the expression data that was filtered out by `selprobes`. A difference

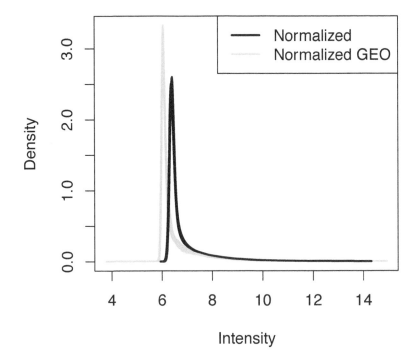

FIGURE 7.2
Distribution of probe intensities for each subject in the expression data set
GSE63061. In grey, the normalized distributions as reported in GEO and in
black the normalized distributions as obtained here.

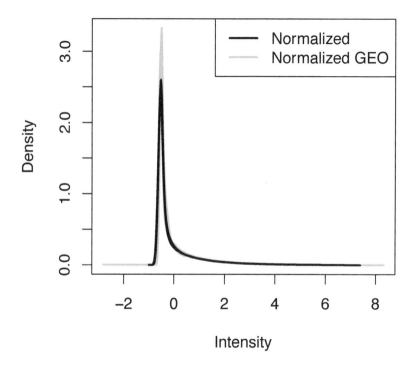

FIGURE 7.3
Distribution of probe intensities for each subject in the expression data set
GSE63061 displaced to zero mean. In grey, the normalized distributions as
reported in GEO and in black the normalized distributions as obtained here.

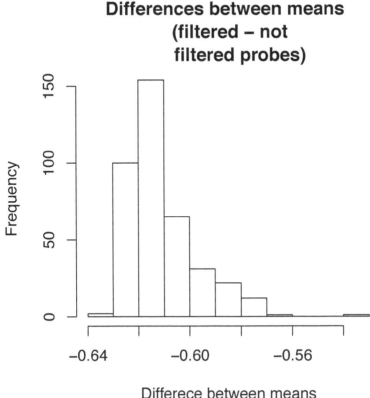

FIGURE 7.4
Histogram for the difference of hybridization means between filtered-in probes
and filtered-out probes across subjects.

between the expression means of the probes across subjects indicates that the
filtered-out probes consistently hybridized less than the filtered-in probes

```
> exprNorm.notsel<-exprNorm[!selprobes,]
> expr.notsel<-exprs(exprNorm.notsel)
>
> hist(colMeans(expr.notsel)-colMeans(expr),
+       main="Differences between means \n (filtered - not
+       filtered probes)", xlab="Differece between means")
```

Furthermore, we see that 90% of the filtering is explained by the fact that
the 80%-quantile of the filtered-out probes is lower than the median of the
entire transcription across subjects and probes.

```
> expr.all<-exprs(exprNorm)
>
> qnthreshold<-quantile(expr.all,0.5)
> checkquants<-sapply(as.data.frame(t(expr.all)), quantile,0.8)
>
> tb<-table(selprobes, checkquants>qnthreshold)
> tb

selprobes FALSE  TRUE
    FALSE 13147  2127
    TRUE   2933 29116
> sum(diag(tb))/sum(tb)
[1] 0.8930752
```

We thus observe that the main trends in the reported dataset GSE63061, with respect to normalization and filtering, can be readily reproduced from the nonnormalized dataset. Small differences between datasets are observed but are not expected to influence the association test that can be obtained from the re-analysis of this study.

7.2.3 Differential expression

The question that we try to answer with a transcriptome-wide association study is whether there are genes whose expression is significantly different between groups of individuals of different phenotypes. The identification of such genes is expected to give insight into the functional differences between the groups. We show the transcriptome-wide analysis for the GSE63061 study using the Bioconductors *limma* package. We extract the expression and phenotype data from the **ExpressionSet** object **gsm.expr**

```
> #expression data
> exprGEO<-exprs(gsm.expr)
>
> #get phenotype data
> pheno <- pData(phenoData(gsm.expr))
```

The subject health status is encoded in **characteristics_ch1**

```
> status <- pheno$characteristics_ch1
> status <- gsub("status: ","", as.character(status))
> fstatus <- factor(status)
> levels(fstatus)<-gsub(" ", "", levels(fstatus))
>
> table(fstatus)
fstatus
          AD borderlineMCI            CTL       CTLtoAD           MCI
         139             3            134             1           109
     MCItoCTL         OTHER
            1             1
```

The study has different groups of individuals. We are interested in finding transcriptomic differences among three groups: AD (Alzheimer's disease), MCI (mild cognitive impairment) and CLT (control).

While two main covariates are age and sex encoded in:
`characteristics_ch1.2` and `characteristics_ch1.3`

```
> age <- substr(pheno$characteristics_ch1.2, 6,7)
> age<-as.numeric(age)
> sex <- pheno$characteristics_ch1.3
```

We also need to extract the surrogate variables that correspond to the model we want to fit, see Chapter 6.

```
> library(sva)
> phenodat<-data.frame(fstatus, age, sex)
>
> mod0 <- model.matrix( ~ age+sex, data = phenodat)
> mod <- model.matrix( ~ fstatus+age+sex, data = phenodat)
>
> svobj <- sva(exprGEO, mod, mod0, n.sv=2)
Number of significant surrogate variables is:  2
Iteration (out of 5 ):1  2  3  4  5
```

A design matrix for the association tests is defined, containing the effect of interest as a factor `fstatus` and the covariates sex, age and the two surrogate variables

```
> sv1 <- svobj$sv[,1]
> sv2 <- svobj$sv[,1]
> design <- model.matrix(~ 0 + fstatus + sex + age + sv1 + sv2)
> colnames(design) <- c(levels(fstatus),"age","sex", "sva1","sva2")
```

The inference in *limma* is performed in two steps. First, a linear fit with the function `lmFit` is made for each gene given the expression data, for which the contrasts of interest are defined; in our case, they are AD vs CTL, MCI vs CTL, AD vs MCI. Then `eBayes` computes the statistics defined by the model.

```
> fit <- lmFit(exprGEO, design)
Coefficients not estimable: sva2
>
> contrast.matrix <- makeContrasts(AD-CTL, MCI-CTL, AD-MCI, levels=design)
>
> fit2 <- contrasts.fit(fit, contrast.matrix)
> fit2 <- eBayes(fit2)
```

Note that a fit for a single contrast, such as one corresponding to a dichotomous status variable could be defined using the corresponding equation model. All significant results across all contrasts are summarized in the Venn diagram shown in figure 7.5 that can be obtained by simply executing:

```
> results <- decideTests(fit2)
> vennDiagram(results, cex=c(1, 0.7, 0.5))
```

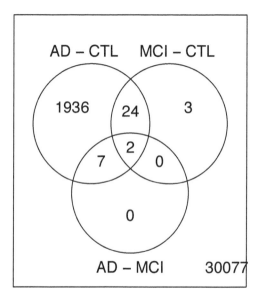

FIGURE 7.5
Venn diagram indicating the number of genes that are differentially expressed
for each comparison in the Alzheimer's disease example.

We can see in the figure 7.5 that most probes with significant results are specific of the AD vs CTL contrast while two genes are significant across all contrast. The object `fit2` contains all the comparisons that are encoded in the different coefficients:

```
> colnames(fit2$coef)
[1] "AD - CTL" "MCI - CTL" "AD - MCI"
```

We observe that the comparison between AD and CTL individuals are in the first column (e.g. first coefficient). Differentially expressed genes between AD and CTL can be obtained using the `topTable` function. Gene symbol can also be added to the final table as following:

```
> tt<-topTable(fit2, coef=1, adjust="BH")
>
> genesIDs <- as.character(fData(gsm.expr)$ILMN_Gene)
> names(genesIDs)<-rownames(gsm.expr)
>
> data.frame(genesIDsgenes=genesIDs[rownames(tt)],
+            logFC=tt$logFC, pvalAdj=tt$adj.P.Val)
              genesIDsgenes       logFC      pvalAdj
ILMN_2189936        RPL36AL  -0.4038645 5.276237e-15
ILMN_1792528       LOC401206  -0.3921189 5.975905e-15
ILMN_1746516          RPS25  -0.3891291 3.765862e-14
ILMN_2189933        RPL36AL  -0.4077166 6.148988e-13
ILMN_1776104         NDUFS5  -0.3678292 6.328579e-13
ILMN_1784286         NDUFA1  -0.4433074 6.328579e-13
ILMN_1652073       LOC653658  -0.4600969 4.704975e-12
ILMN_1732328       LOC646200  -0.4039218 3.182731e-11
ILMN_2097421         MRPL51  -0.2498108 3.182731e-11
ILMN_2232936          UQCRH  -0.3352926 5.137196e-11
```

A volcano plot showing the P-values of association against the log_2 fold change is used to visualize significant results.

```
> volcanoplot(fit2, coef = 1, highlight=5, names=genesIDs, cex=0.2)
```

We thus see that genes with significant expression differences are down-regulated in AD with respect to controls at about -0.4 log_2 fold change.

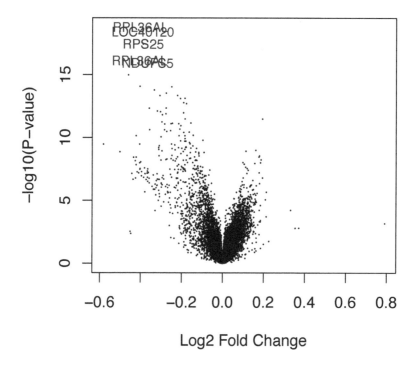

FIGURE 7.6
Volcano plot showing the P-values of association against the log2 fold expression change.

TABLE 7.1
Hypothetical example showing the table of counts obtained from an RNA-seq experiment measured in 6 samples from two conditions.

	Condition A			Condition B		
Gene$_1$	0	0	4	12	14	13
Gene$_2$	0	21	40	45	23	0
Gene$_3$	0	0	4	12	14	13
...	...	...	...	...	...	...
Gene$_G$	234	123	40	125	49	134

7.3 Next generation sequencing data

RNA-seq is a high-throughput technology that scans the transcriptome by sequencing the tRNA content of a biological sample. RNA is randomly cut into small fragments call reads, which are sequenced and mapped to a reference genome . The transcription profile of a particular gene follows from counting the number of times the transcripts of the genes were mapped by sequenced reads. The summarized RNA-seq data is known as count data. Figure 7.3 describes the main steps that are taken to obtain the count table that measures the transcription of active genes in biological samples. The sequenced reads can be counted in a number of different ways:

- By alignment to the genome and summarized at either gene or transcript (isoform) level.

- By alignment to the transcriptome and summarized at either gene or transcript (isoform) level.

- By assembling directly into transcripts and summarized at either gene or transcript (isoform) level.

In this section, we explain how counts at the gene-level can be used for differential expression profiling to help explain phenotypic differences between subjects. Important studies that measure the differences in splicing and isoform ratios between groups are not covered. Table 7.1 shows a hypothetical RNA-seq data of G genes obtained for 6 samples, belonging to two different conditions. The main goal is to discover the genes that are differentially expressed between individuals from condition A and B.

In RNA-seq analysis, we deal with the number of reads (counts) that map to the biological feature of interest (gene, transcript, exon, etc.). The count number depends linearly with the abundance of the target's transcription because the sequencing of RNA is a direct measure of transcription . This is considered as one of the advantages over microarrays that indirectly measure

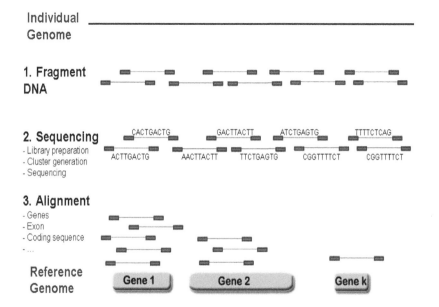

FIGURE 7.7
RNA-seq scheme to get count data.

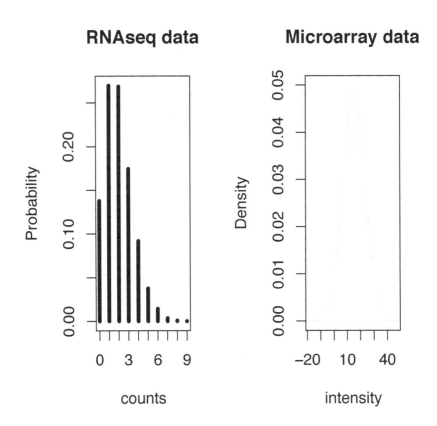

FIGURE 7.8
Gene expression distribution obtained from RNA-seq and microarray data from a hypothetical gene.

transcription by hybridization. Figure 7.8 illustrates the type of data from microarray and RNA-seq experiments. While microarray data is continuous, RNA-seq is discrete and, therefore, the modeling of each type of data is different.

There are two important factors that influence the number of gene counts and which need to be taken into account, see figure 7.3. The first factor is the sequencing depth or library size, that is, the total number of reads mapped to the genome; the second factor is the gene length, i.e. the number of bases covering a gene. It is expected that larger genes, for a given level of transcription, will have more gene counts.

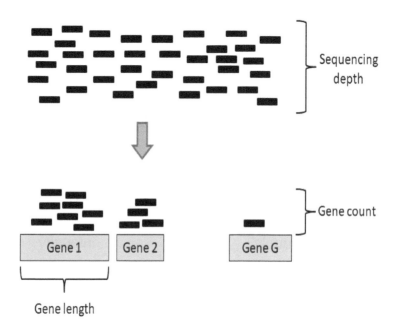

FIGURE 7.9
Key concepts involved in an RNA-seq experiment.

7.3.1 Normalization

Count data must be normalized across genes since the number of counts that map to a given gene depends on sequencing depth (e.g library size), gene length and GC-content. The aim of normalization is also to remove the systematic technical effects to ensure that they have minimal impact on transcription profiling [125].

There are several methods to normalize RNA-seq data which take into account different issues.

- **RPKM** [98]: Counts are divided by the transcript length (kb) times the total number of millions of mapped reads, thus, controlling for sequencing depth and gene length.

$$\text{RPKM} = \frac{\dfrac{\text{number of reads in region}}{\text{region length} \times 10^3}}{\text{total reads} \times 10^6}$$

- **TMM** [125]: Trimmed mean of M values. This method tries to improve RPKM by considering not only the length of the gene but also the whole distribution across the genome.

- **EDAseq** [98]: Within-lane gene-level GC-content normalization. This method corrects for library size, gene length and GC-content.

- **CQN** [50]: Conditional quantile normalization (CQN) algorithm combining robust generalized regression. This method corrects for library size, gene length and GC-content.

- **Others**: Upper-quartile [13]; FPKM [158]: Instead of counts, `Cufflinks` software generates FPKM values (Fragments Per Kilobase of exon per Million fragments mapped) to estimate gene expression, which are analogous to RPKM [158].

Normalization procedures are an active field of research. The previous methods are the most common and have been comprehensively compared against each other and with others in different studies [141, 83].

We demonstrate different methods to normalize the count data from lymphoblastoid cell lines from 69 unrelated Nigerian individuals, described in [110]. This data are available through the Bioconductors *tweeDEseqCountData* package.

```
> library(tweeDEseqCountData)
> data(pickrell)
> pickrell.eset
ExpressionSet (storageMode: lockedEnvironment)
assayData: 52580 features, 69 samples
  element names: exprs
protocolData: none
```

```
phenoData
  sampleNames: NA18486 NA18498 ... NA19257 (69 total)
  varLabels: num.tech.reps population study gender
  varMetadata: labelDescription
featureData
  featureNames: ENSG00000000003 ENSG00000000005 ... LRG_99 (52580
    total)
  fvarLabels: gene
  fvarMetadata: labelDescription
experimentData: use 'experimentData(object)'
Annotation:
```

The object `pickrell.est` is an **ExpressionSet** containing RNA-seq count data obtained from the ReCount repository available at `http://bowtie-bio` `.sourceforge.net/recount`. Details on the pre-processing steps to obtain this table of counts from the raw reads of [110] are provided on the website and in Frazee et al. [37].

Our aim is to use this data to compare different normalization methods. Perhaps, the most simple normalization is to standardize the counts dividing them by the total number of reads in the sample (size of the libraries):

```
> counts <- exprs(pickrell.eset)
> lib.size <- colSums(counts)
> NormByTotalNrReads <- sweep(counts, 2, FUN="/", lib.size)
```

For an RPKM (reads per kilobase of transcript and million mapped reads) normalization, we first need to know the length of query regions (e.g. length of genes). This information is also available from the *tweeDEseqCountData* package, but can also be obtained from *biomaRt* (introduced in Section 4.3.1).

```
> data(annotEnsembl63)
> head(annotEnsembl63)
                 Symbol Chr    Start      End EntrezID
ENSG00000252775     U7   5 133913821 133913880     <NA>
ENSG00000207459     U6   5 133970529 133970635     <NA>
ENSG00000252899     U7   5 133997420 133997479     <NA>
ENSG00000201298     U6   5 134036862 134036968     <NA>
ENSG00000222266     U6   5 134051173 134051272     <NA>
ENSG00000222924     U6   5 137405044 137405147     <NA>
                                     Description Length
ENSG00000252775 U7 small nuclear RNA [Source:RFAM;Acc:RF00066]     NA
ENSG00000207459 U6 spliceosomal RNA  [Source:RFAM;Acc:RF00026]     NA
ENSG00000252899 U7 small nuclear RNA [Source:RFAM;Acc:RF00066]     NA
ENSG00000201298 U6 spliceosomal RNA  [Source:RFAM;Acc:RF00026]     NA
ENSG00000222266 U6 spliceosomal RNA  [Source:RFAM;Acc:RF00026]     NA
ENSG00000222924 U6 spliceosomal RNA  [Source:RFAM;Acc:RF00026]     NA
                GCcontent
ENSG00000252775        NA
ENSG00000207459        NA
ENSG00000252899        NA
ENSG00000201298        NA
ENSG00000222266        NA
ENSG00000222924        NA
```

```
> genes.ok <- intersect(as.character(rownames(counts)),
+                        as.character(rownames(annotEnsembl63)))
> geneAnnot <- annotEnsembl63[genes.ok,]
> counts.ok <- counts[genes.ok,]
> identical(rownames(geneAnnot), rownames(counts.ok))
[1] TRUE
```

We obtain annotation data in **annotEnsembl63** and map it to **counts**, we then check that both *data.frames* are in the same order. The RPKM normalization is computed by

```
> width <- geneAnnot$Length
> NormByRPKM <- t(t(counts.ok / width *1000)
+                 /colSums(counts.ok)*1e6)
```

We now apply a TMM (trimmed mean of M-values) normalization, which was developed to correct the gene counts by the expression properties of the whole sample[125]. The TMM normalization method is implemented in the Bioconductors **tweeDEseq** package and can be performed by:

```
> library(tweeDEseq)
> NormByTMM <- normalizeCounts(counts.ok, method="TMM")
```

Another type of normalization is CQN, available in the package *cqn*. It requires different steps that have been encapsulated in the function called **normalizeCounts** from *tweeDEseq*. The function requires the length and the percentage of GC-content of each gene.

```
> library(cqn)
> annotation <- geneAnnot[,c("Length", "GCcontent")]
> NormByCQN <- normalizeCounts(counts.ok, method="cqn",
+                              annot=annotation)
RQ fit ...............................................................
SQN .
```

We now compare the performance of each normalization method. Assuming that between two samples, most genes are not differentially expressed, the distribution of the difference of log-ratios between the samples should be centered around 0 when data is correctly normalized. We thus examine the distributions for the 1st and 2nd sample under normalizations by the total number of reads and by RPKM (Figure 7.10):

```
> MbyT <- log2(NormByTotalNrReads[, 1] / NormByTotalNrReads[, 2])
> MbyRPKM <- log2(NormByRPKM[, 1] / NormByRPKM[, 2])
>
> par(mfrow=c(1,2))
> hist(MbyT, xlab="log2-ratio", main="Total reads")
> abline(v=0, col="red")
> hist(MbyRPKM, xlab="log2-ratio", main="RPKM")
> abline(v=0, col="red")
```

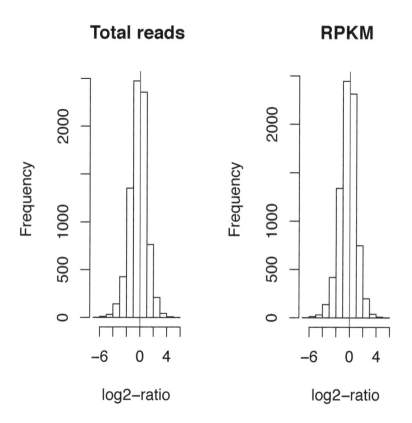

FIGURE 7.10
Comparison of log-ratio count intensity of samples 1 and 2 from the Pickrell
dataset.

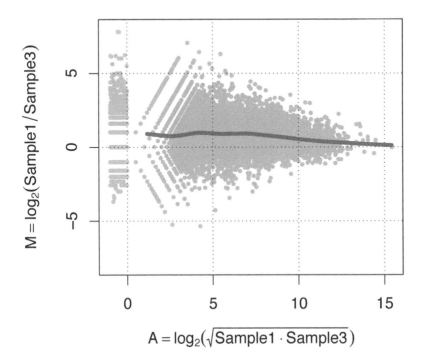

FIGURE 7.11
MA-plot of raw (e.g. non-normalized) Pickrell data on samples 1 and 3.

Figure 7.10, however, does not show whether the null difference between samples holds for different levels of gene expression. With a MA-plot, one can check whether data has been correctly normalized at any expression level. The MA-plot is available from the Bioconductor package *edgeR*. We first illustrate the MA-plot for nonnormalized counts of samples 1 and 2, in figure 7.11.

```
> library(edgeR)
> maPlot(counts[,1], counts[,2], pch=19, cex=.5, ylim=c(-8,8),
+       allCol="darkgray", lowess=TRUE,
+       xlab=expression( A == log[2] (sqrt(Sample1 %.% Sample3)) ),
+       ylab=expression(M == log[2](Sample1/Sample3)))
>       grid(col="black")
```

The resulting plot can be seen in Figure 7.12 (top left part). In the X-axis the plot shows the mean between the log-ratios. In the Y-axis the plot

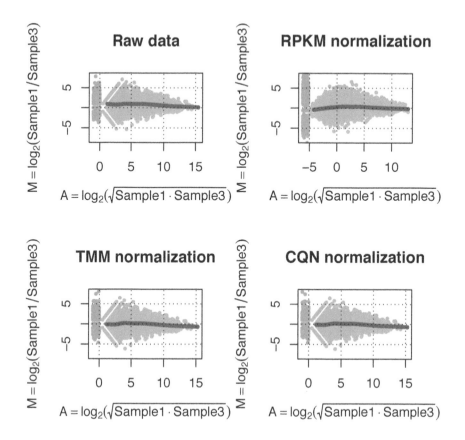

FIGURE 7.12
MA-plot of Pickrell data on samples 1 and 3 for raw data and normalized data
using RPKM, TMM and CQN methods.

shows the difference of log-ratios between the samples, similar to figure 7.10.
The MA-plot is, therefore, the difference against the mean of the gene counts
between the two samples. The red line shows the expected M-values as a
function of A-values. In our example of nonnormalized data, we can see that
for genes with low counts the distribution of the difference between samples is
not zero; in particular, we observe that sample 2 has more counts than sample
1.

We now compare the normalization methods using MA-plots, replacing
`counts` by `NormByRPKM`, `NormByTMM` and `NormByTMM`, respectively.

Figure 7.12 shows, for instance, that TMM and CQN may be preferable,
as the expectation of Y is close to 0 across different levels of gene expression. When comparing gene expression between conditions, MA-plots can be

computed for each condition to verify that there are no systematic differences between them.

7.3.2   Gene filtering

Not all genes in a biological sample are expected to be active and, therefore, one can filter genes with low expression to improve statistical inference. There is not a general rule about filtering but removing genes with less than 5 counts per million in all samples is a common practice. We filter the data in our example with the `filterCounts` from *tweeDEseq*

```
> NormByCQN.f <- filterCounts(NormByCQN)
> dim(NormByCQN)
[1] 40279    69
> dim(NormByCQN.f)
[1] 40272    69
```

With the argument `mean.cpm.cutoff` one can set the minimum mean expression level through all the samples.

7.3.3   Differential expression

RNA-seq data are discrete therefore linear models, such that those implemented in *limma* cannot be applied. While it is common practice to apply logarithmic transformations before fitting linear models, the transformations depend on an offset level to account for zero counts, which, in turn, can affect the group differences assessed in regression models. Other transformations may be applied, however, it is more adequate to make inferences based on distributions of count data such as the Poisson or the negative binomial distributions [91, 124, 2]. Negative binomial modeling is preferred over Poisson's, as biological variability results in a difference between the mean and variance of the data. The negative binomial is defined by the parameters λ and ϕ that model the intensity and overdispersion of data.

Let us then assume that X_{gA} corresponds to the number of reads that mapped into gene g ($g = 1, \ldots, G$) across subjects within condition A and that $X_{gA} \sim NB(\lambda_{gA}, \phi_g)$. If counts in condition B also distribute $X_{gB} \sim NB(\lambda_{gB}, \phi_g)$ then we aim to test whether λ_{gA} and λ_{gB} are significantly different across all g. The dispersion of the gene ϕ_g cannot be estimated when few individuals are analyzed, therefore we can:

- Estimate a common dispersion ϕ for all g using the conditional likelihood approach proposed by [124]. This method is implemented in the Bioconductors package *edgeR*, or

- Estimate ϕ_g as a function of the mean [2], as implemented in the Bioconductors package *DESeq*.

We demonstrate how to perform both estimations of the dispersion parameter, using our previous example of lymphoblastoid cell lines. Our final aim is to detect the genes that are differentially expressed between males and females.

From the CQN normalized counts, we first estimate the common dispersion approach using the *edgeR* package

```
> library(edgeR)
> sex <- pData(pickrell.eset)$gender
> d <- DGEList(counts = NormByCQN, group = as.factor(sex))
> d <- estimateCommonDisp(d)
> d$common.dispersion
[1] 0.1214059
```

which corresponds to little extra variability between technical replicates

```
> sqrt(d$common.dispersion)
[1] 0.3484335
```

edgeR also allows an estimation of ϕ_g for each gene, which, after estimating the common dispersion, is obtained by the `estimateTagwiseDisp`

```
> d <- estimateTagwiseDisp(d)
> names(d)
 [1] "counts"          "samples"           "common.dispersion"
 [4] "pseudo.counts"   "pseudo.lib.size"   "AveLogCPM"
 [7] "prior.df"        "prior.n"           "tagwise.dispersion"
[10] "span"
```

Dispersion estimates with *DESeq* first requires the creation of a `CountDataSet` object, on which it estimates the library size (e.g. size factors) for the gene dispersion

```
> library(DESeq)
> cds <- newCountDataSet(NormByCQN, as.factor(sex))
> cds <- estimateSizeFactors(cds)
> cds <- estimateDispersions(cds)
> cds
CountDataSet (storageMode: environment)
assayData: 40279 features, 69 samples
  element names: counts
protocolData: none
phenoData
  sampleNames: NA18486 NA18498 ... NA19257 (69 total)
  varLabels: sizeFactor condition
  varMetadata: labelDescription
featureData
  featureNames: ENSG00000000003 ENSG00000000005 ...
    ENSG00000254396 (40279 total)
  fvarLabels: disp_pooled
  fvarMetadata: labelDescription
experimentData: use 'experimentData(object)'
Annotation:
```

From the dispersion estimates, we are then interested in testing the null hypotheses

$$H_0 : \lambda_{gA} = \lambda_{gB}$$

for $g = 1, \ldots, G$ and A are females and B are males.

The Wald test, the score test or the likelihood ratio test are standard options for hypothesis testing for the differences in the mean of two negative binomial distributions. When analyzing small sample sizes, exact tests are recommended. The `exactTest` function from *edgeR* can be used with common and tagwise dispersion estimates

```
> resEdgeR.common <- exactTest(d, pair=c("female", "male"),
+                              dispersion="common")
> resEdgeR.tagwise <- exactTest(d, pair=c("female", "male"),
+                              dispersion="tagwise")
```

The functions allow the user to indicate the reference category to facilitate interpretation of the results (Negative vs Positive). Inferences by *edgeR* can be easily visualized

```
> topTags(resEdgeR.common)
Comparison of groups:  male-female
                   logFC    logCPM         PValue           FDR
ENSG00000129824  7.159242  6.431781  0.000000e+00  0.000000e+00
ENSG00000099749  4.367018  3.873497  2.152635e-156 4.335298e-152
ENSG00000154620  3.654947  2.999080  8.303703e-99  1.114883e-94
ENSG00000198692  3.469197  3.178589  2.797641e-97  2.817154e-93
ENSG00000157828  3.144968  2.061472  1.219428e-58  9.823468e-55
ENSG00000082397  2.270122  2.616552  3.582617e-43  2.405071e-39
ENSG00000008196  1.925912  2.195954  9.880004e-28  5.685096e-24
ENSG00000173110 -1.462254  4.598439  2.294780e-25  1.155393e-21
ENSG00000240563  1.574258  3.030833  1.398688e-24  6.071588e-21
ENSG00000172543  1.510126  3.338356  1.507383e-24  6.071588e-21


> topTags(resEdgeR.tagwise)
Comparison of groups:  male-female
                   logFC     logCPM         PValue          FDR
ENSG00000129824  7.1666244  6.431781  0.000000e+00  0.000000e+00
ENSG00000099749  4.3674469  3.873497  8.933988e-161 1.799260e-156
ENSG00000198692  3.4689855  3.178589  2.184359e-104 2.932793e-100
ENSG00000154620  3.6551601  2.999080  3.381247e-103 3.404832e-99
ENSG00000157828  3.1712920  2.061472  1.378954e-38  1.110858e-34
ENSG00000006757 -0.9876989  6.014201  9.330792e-15  6.263916e-11
ENSG00000082397  2.2650505  2.616552  1.212797e-08  6.978607e-05
ENSG00000008196  1.8734061  2.195954  1.165919e-06  5.870257e-03
ENSG00000101846 -0.8425243  6.995108  1.630358e-06  7.296575e-03
ENSG00000086712 -0.4928693  5.565841  3.928195e-06  1.582238e-02
```

On the other hand, `nbinomTest` from *DESeq* computes the hypothesis testing

```
> resDESeq <- nbinomTest(cds, "female", "male")
```

that can be summarized by

```
> out <- resDESeq[resDESeq$padj < 0.05,]
> out.o <- out[order(out$padj),]
> head(out.o)
                     id  baseMean  baseMeanA baseMeanB  foldChange
6249  ENSG00000129824 192.43755   3.016056 453.70858 150.4311094
2133  ENSG00000099749  31.30968   2.994394  70.36524  23.4989931
9910  ENSG00000154620  15.76266   2.597906  33.92093  13.0570261
18252 ENSG00000198692  18.82180   3.206867  40.35964  12.5853828
10276 ENSG00000157828   7.89314   1.522690  16.67997  10.9542756
163   ENSG00000006757 140.01558 175.738495  90.74259   0.5163501
      log2FoldChange          pval          padj
6249       7.2329591 1.532607e-236 6.173187e-232
2133       4.5545270  1.720723e-60  3.465449e-56
9910       3.7067544  1.128007e-57  1.514500e-53
18252      3.6536772  3.762231e-47  3.788473e-43
10276      3.4534222  3.031194e-20  2.441869e-16
163       -0.9535785  2.479614e-07  1.664606e-03
```

Differential expression analysis typically requires adjusting for covariates that can act as confounders, such as batch effect, gender, age, cell count or non-observed variables (e.g. those obtained from surrogate variable analysis, see Chapter 6). In those cases, exact methods are not appropriate and generalized linear models should be used instead. The package *DESeq2* is the extension of *DESEq* that includes functions to adjust for covariates. *edgeR* also has functions for adjustment, including those that may be derived by *voom* [75], which transforms the counts into continuous variables. For these continuous variables, the transcriptomic-wide association analysis with the *limma* package, as described for microarray data, can be retaken.

We illustrate the use of the *DESeq2* package with our example of transcriptomic sex differences. *DESeq2* uses nonnormalized data, as normalization is internally implemented. Comparisons between *DESeq2* normalization and the methods described in the previous section have been done by Reddy [119]. For assessing transcriptomic differences between conditions with *DESeq2*, we first implement the model using the **DESeqDataSetFromMatrix** function

```
> library(DESeq2)
> counts <- exprs(pickrell.eset)
> pheno <- pData(pickrell.eset)
> dds <- DESeqDataSetFromMatrix(countData = counts,
+              colData = pheno,
+              design = ~ gender)
```

Notice the reference category

```
> table(pheno$gender)

female   male
    40     29
```

which can be changed by the `relevel` function if needed. Models that adjust by covariates are specified in the `design` argument (e.g. `design =    gender + covar1 + ... + covar2`). *DESeq2* allows filtering genes with low number of counts. Strict filtering to increase power can be applied via independent filtering [12]. While the procedure is implemented in *DESeq2* through the `results` function, here we directly filter the data

```
> keep <- rowSums(counts(dds)) >= 10
> dds <- dds[keep,]
```

where genes with expression counts over 10 are to be filtered in. Inferences of differential expression are then performed by applying the function `DESeq` on the previously defined model.

```
> dds <- DESeq(dds)
```

We can see that the `DESeq` function performs all the required steps to analyze RNA-seq data. The function estimates size factors and dispersion parameters and performs the differential expression analysis. Results can be retrieved by `results`

```
> resDESeq2  <- results(dds, pAdjustMethod = "fdr")
> head(resDESeq2)
log2 fold change (MLE): gender male vs female
Wald test p-value: gender male vs female
DataFrame with 6 rows and 6 columns
```

	baseMean	log2FoldChange	lfcSE
	<numeric>	<numeric>	<numeric>
ENSG00000000419	62.7163896330933	0.0400414411623122	0.113501829926241
ENSG00000000457	70.4681652433405	0.0295898365508509	0.111753599826055
ENSG00000000460	13.00098412767	0.164860196697704	0.151014657187751
ENSG00000000938	43.5641298200903	0.114079034890195	0.195372881677187
ENSG00000001036	37.9519065843039	-0.0059235513995029	0.10791382594948
ENSG00000001167	228.480251542072	0.05203096772429	0.0966825209508479

	stat	pvalue	padj
	<numeric>	<numeric>	<numeric>
ENSG00000000419	0.352782340058596	0.724251624225412	0.97102721328611
ENSG00000000457	0.264777480071403	0.791180886203641	0.981324127233597
ENSG00000000460	1.09168341515843	0.274972277379928	0.820696892911289
ENSG00000000938	0.583904142227307	0.559284806167601	0.945493135010185
ENSG00000001036	-0.054891496500884	0.956224906475762	0.997304774663576
ENSG00000001167	0.538163126205014	0.590464434492261	0.953104643286861

The results table can be sorted by level of significance

```
> resDESeq2 <- resDESeq2[order(resDESeq2$padj), ]
> resDESeq2
log2 fold change (MLE): gender male vs female
Wald test p-value: gender male vs female
DataFrame with 10151 rows and 6 columns
```

	baseMean	log2FoldChange	lfcSE
	<numeric>	<numeric>	<numeric>

```
ENSG00000129824  153.697416199909       10.3899345177552  0.339922357975214
ENSG00000099749  15.3584714038195        7.58112234334661  0.346299643964625
ENSG00000198692  8.19446402737211        6.79861905352584  0.368919916989441
ENSG00000154620  10.901220933584         5.88749486686232  0.344180258962215
ENSG00000157828  5.71039799565385        6.10942122446629  0.459802604726422

...                       ...                    ...                    ...
ENSG00000151176  228.864854705402  -1.21436602770726e-05  0.108178624066344
ENSG00000157833  0.749251006983084  0.000463087528811732  0.537262141782636
ENSG00000203993  93.449256236496    1.82294654969646e-05  0.0907627462697629
ENSG00000204791  1.59867468118867  -5.69438895737244e-05  0.394238803766014
ENSG00000204019                0                      0                 0
                                           stat              pvalue
                                      <numeric>           <numeric>
ENSG00000129824         30.5656108637396  3.50757846993193e-205
ENSG00000099749         21.8917994155403  3.10992855498796e-106
ENSG00000198692         18.4284413511901   7.76892648025084e-76
ENSG00000154620         17.1058470483301    1.3423595633559e-65
ENSG00000157828         13.2870522299484   2.75208179715074e-40

...                            ...                    ...
ENSG00000151176  -0.000112255636285641        0.999910432961133
ENSG00000157833   0.00086193962462199         0.999312271766327
ENSG00000203993   0.000200847442879079        0.999839746927327
ENSG00000204791  -0.000144440093237299        0.999884753480046
ENSG00000204019                       0                        1
                                      padj
                                 <numeric>
ENSG00000129824  3.56019214698091e-201
ENSG00000099749  1.57828874165639e-102
ENSG00000198692   2.62848679248487e-72
ENSG00000154620   3.40623739201559e-62
ENSG00000157828    5.586726048216e-37

...                            ...
ENSG00000151176       0.999910432961133
ENSG00000157833       0.999910432961133
ENSG00000203993       0.999910432961133
ENSG00000204791       0.999910432961133
ENSG00000204019                      NA
```

We can obtain the MA-plot, see Figure (7.13), between conditions to check the internal normalization of data

```
> plotMA(resDESeq2)
```

We can also produce a boxplot that compares the count differences between males and females for the top gene (Figure 7.14)

```
> plotCounts(dds, gene = "ENSG00000129824",
+             intgroup = "gender")
```

Transcriptomic profiling can also be performed using the voom method implemented in the *limma* package. This method estimates the mean-variance relationship of the log-counts, generates a precision weight for each observation and uses them in an empirical Bayes analysis pipeline. As such, it allows the use of linear models to analyze read counts from RNA-seq experiments [75].

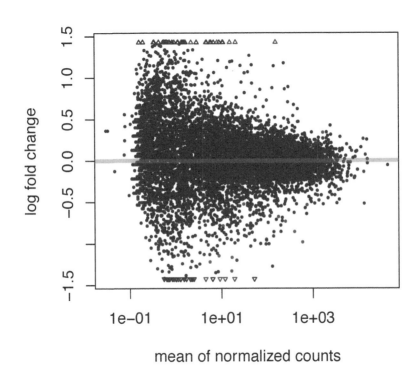

FIGURE 7.13
MA-plot comparing both conditions of Pickrell data.

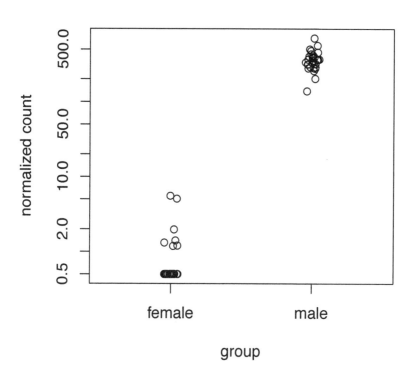

FIGURE 7.14
Gene expression levels of females and males corresponding to the gene most associated with gender status in the Pickrell dataset.

voom: Mean–variance trend

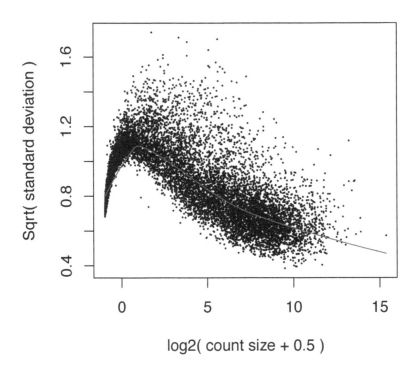

FIGURE 7.15
Mean-variance relationsip corresponding to the Pickrell dataset.

For analyzing our example with this approach, we first apply the voom function to count data:

```
> library(limma)
> design <- model.matrix( ~ gender, data=pheno)
> v <- voom(counts, design=design, plot=TRUE)
```

Figure 7.15 shows the mean-variance relationship, from which precision weights are given to the count data, so continuous data are derived and analyzed by the usual *limma* procedure

```
> fit <- lmFit(v, design)
> fit <- eBayes(fit)
> topTable(fit, coef=ncol(design))
                     logFC      AveExpr        t      P.Value    adj.P.Val
```

```
ENSG00000129824   9.1311273   1.95469533  54.225084  1.065179e-64  5.600714e-60
ENSG00000099749   6.1321934   0.40983758  45.639189  6.485999e-59  1.705169e-54
ENSG00000198692   5.1906429  -0.03456802  35.287216  2.001501e-50  3.507964e-46
ENSG00000154620   4.9963042   0.48128403  25.017166  1.672057e-39  2.197919e-35
ENSG00000157828   4.1312237  -0.38999476  16.850518  5.322729e-28  5.597382e-24
ENSG00000006757  -0.9164784   5.30180979  -9.994009  1.027507e-15  7.718044e-12
ENSG00000183878   1.8245692  -1.50072787  10.254164  3.210781e-16  2.813714e-12
ENSG00000092377   0.7785641  -1.95781202   4.607544  1.524213e-05  8.974261e-02
ENSG00000177606  -0.5767409   8.24149085  -4.269167  5.368351e-05  8.974261e-02
ENSG00000143921   0.8071775  -1.86105437   4.513827  2.170952e-05  8.974261e-02
                            B
ENSG00000129824  67.155001
ENSG00000099749  56.640302
ENSG00000198692  48.692481
ENSG00000154620  41.492680
ENSG00000157828  30.056641
ENSG00000006757  20.728249
ENSG00000183878  15.860830
ENSG00000092377   1.957700
ENSG00000177606   1.735214
ENSG00000143921   1.623161
```

We see that the top genes with significant differences of transcription between males and females are consistent with those found using the *DESeq2*.

8

Epigenomic studies

CONTENTS

8.1 Chapter overview

In this chapter, we analyze epigenomic data in the form of DNA methylation. Our aim is to identify sites where differences in methylation can help to explain phenotypic differences between individuals. Numerous methylomic datasets that are publicly available are from microarray experiments. Many methods for preprocessing data and performing association analysis are similar to those explained for transcriptomic data. However, important differences need to be taken into account, in particular, methylation of neighboring sites are correlated and, therefore, various approaches to test the association between methylomic data and phenotypic differences can be used. Some of the most common approaches are available through the Bioconductor package *MEAL*. Here, we demonstrate the use of the package on the GEO study of Alzheimer's disease GSE80970, introduced in Chapter 2. We show how to perform different analyses within a single pipeline and how to extract, evaluate and plot results.

8.2 Epigenome-wide association studies

The accessibility of DNA is essential to determine the level of gene transcription. The addition of a methyl group at the CpG sites in the coding, promoter or transcription factor binding regions can impair accessibility to the gene and affect its transcription. Methylation is a complex process that can be determined by environmental factors, and by developmental or tissue-specific processes. Methylation is inherited by daughter cells and therefore constitute one of the mechanisms for epigenomic (nongenetic) transmission. Therefore, identification of methylated sites can help to identify genes that play important roles in physiology or disease that may be coupled with environmental conditions. As such, the main aim of epigenome-wide association studies (EWAS) is to identify methylation sites that are significantly associated with a phenotype of interest.

Methylation occurs in the CpC sites, which constitute 1% of the genome. The methylation status of the CpG sites across a genome can be inquired with microarrays, where DNA at the cytosine site of the CpG sequence is first transformed to thymine if it is methylated, and then hybridized to array probes with C or T genotypes, corresponding to unmethylated or methylated probes. Therefore, most of the analyses that we have discussed for SNP and transcription microarrays apply. Specific differences arise through the differences in the underlying biological processes. For instance, methylation at a CpG is a discrete state for a given chromosome sequence of a single cell. However, not all the cells in a biological sample are methylated and, therefore, the observed methylation average for a biological sample distributes continuously between 0 and 1, following a beta distribution. In addition, methylation is radically different between tissues, and therefore when having cell-type mixture is important to adjust by the cellular mixture ratios in association analyses.

8.3   Methylation arrays

Illumina Infinium HumanMethylation 450K BeadChip assay is a common tool to sample the methylome in human samples. Public methylomic data is widely available in GEO or projects like TCGA. Their 450,000 probes cover most of the methylation states of the human genome, being one of the reasons for its success. Most publicly available data have been normalized and pre-processed, and raw hybridizing intensities have been converted into normalized methylation ratios. Bioconductor packages such as *minfi* [4] or *lumi* [30] have functions for preprocessing and inferring differentially methylated probes (DMPs) and differentially methylated regions (DMRs). However, as there are multiple approaches that also take into account the correlation between methylation sites, it is convenient to perform different types of association analysis on methylomic data. Many of these methods are available in different software packages.

8.4   Differential methylation analysis

MEAL Bioconductor package aims to facilitate the analysis of methylation arrays, such as Illumina Methylation 450K, by integrating different pipelines from different packages. It includes two methods to analyze DMPs by comparing means or variances across conditions, three DMRs detection algorithms (bumphunter, blockFinder and DMRcate) and a method to test differences in methylation in a target region (RDA). *MEAL* also includes functions to visualize the results as well as a wrapper function to perform all the required analyses in an EWAS.

We illustrate how to perform methylomic data analyses using the Alzheimer's study introduced in Chapter 2 available in GEO

```
> library(GEOquery)
> gsm.meth <- getGEO("GSE80970", destdir = ".")
> gsm.meth <- gsm.meth[[1]]
```

MEAL has a range of analysis methods for methylation data that is structured in a *GenomicRatioSet* class, from the *minfi* package. As GEO typically provides data as an *ExpressionSet* object, we start by transforming the data into the required class

```
> library(minfi)
> betas <- exprs(gsm.meth)
> pheno <- pData(gsm.meth)
> gsm.grs <- makeGenomicRatioSetFromMatrix(mat=betas,
+                                          pData=pheno)
```

The 450K Illumina array includes probes that are common SNPs or that are located in sex chromosomes. We filter these probes

```
> gsm.grs <- dropMethylationLoci(gsm.grs)
> gsm.grs <- dropLociWithSnps(gsm.grs)
> rr <- seqnames(rowRanges(gsm.grs))
> gsm.grs.f  <- gsm.grs[!rr%in%c("chrX", "chrY"),]
```

As we want to compare the methylation state between cases and controls across the methylome, we extract the case-control status of individuals from the variable in the phenotype dataset (i.e. **characteristics_ch1.1**)

```
> status <- gsm.grs.f$characteristics_ch1.1
> status <- gsub("disease status: ","", as.character(status))
> gsm.grs.f$status <- as.factor(status)
> table(gsm.grs.f$status)

Alzheimer's disease             control
                148                 138
```

The function `runPipeline` can run different analyses for one dataset: DiffMean, DiffVar, bumphunter, blockFinder, DMRcate and RDA. DiffMean and DiffVar are methods to analyze individual CpGs: DiffMean tests whether there is a significant difference between methylation means between groups. This is the massive univariate testing equivalent to other *omic* association analyses. the DiffVar method tests the variance differences in each CpG between groups. Epigenetic instability or the loss of epigenetic control of important genomic domains can lead to increased methylation variability in some diseases such as cancer [51]. The methylation state of neighboring CpGs is usually correlated. Therefore, methods like bumphunter, blockFinder and DMRcate are multi-site CpG methods that aim to identify regions in which there are congruent significant methylation differences between groups. Whereas, RDA estimates differences in methylation for predefined regions.

To illustrate how *MEAL* performs all these analyses, we first remove missing values in the data since methods like DiffVar require complete cases. We identify missing values with the function `getBeta`, which extracts the mean methyaltion state of probes across the samples.

```
> sum(is.na(getBeta(gsm.grs.f)))
[1] 123992
```

Probes are then removed from the analyses

```
> gsm.grs.noNA <- gsm.grs.f[rowSums(is.na(getBeta(gsm.grs.f))) == 0,]
> nrow(gsm.grs.f)
[1] 453526
> nrow(gsm.grs.noNA)
[1] 380733
```

We therefore observe that the number of CpGs is reduced from 453526 to 380733. Alternatively, data can be imputed using the *impute* Bioconductors package.

To perform the methylation association analyses, we use the function `runPipeline` that is a method for objects of class *GenomicRatioSet*, *ExpressionSet* and *SummarizedExperiment*. The function allows the analysis of more than one response variable (argument `variable_names`) as well as adjustment by covariates (argument `covariable_names`). Further details can be found in MEAL's vignette [130]. By default, `runPipeline` runs all analyses.

```
> library(MEAL)
```

```
> methRes <- runPipeline(set = gsm.grs.noNA,
+                         variable_names = "status")
> methRes
Object of class 'ResultSet'
 . created with: runPipeline
 . sva:  no
 . #results: 5 ( error: 1 )
```

```
. featureData: 380733 probes x 5 variables
> names(methRes)
[1] "DiffMean"     "DiffVar"      "bumphunter"  "blockFinder" "dmrcate"
```

The analysis generates a `ResultSet` object containing the results across six different analyses: DiffMean, DiffVar, bumphunter, blockFinder, DMRcate and RDA. The six methods are based in the same linear model which is specified in the parameters `variable_names` and `covariable_names`. The goodness of fit of the models is also evaluated.

Access to any of these results can be obtained by using the `getAssociation` function and indicating the type of analysis in argument `rid`. This provides the list of DMPs:

```
> head(getAssociation(methRes, rid="DiffMean"))
                logFC        CI.L        CI.R   AveExpr         t
cg08991643 -0.04019226 -0.05270129 -0.02768322 0.3141154 -6.324222
cg00472710 -0.03836907 -0.05068685 -0.02605129 0.2457141 -6.131085
cg14962509 -0.03016704 -0.03992891 -0.02040517 0.7824268 -6.082585
cg22454769 -0.03333790 -0.04440614 -0.02226966 0.4019419 -5.928545
cg13772815 -0.02695197 -0.03595163 -0.01795231 0.4447270 -5.894579
cg00228891 -0.02849172 -0.03813989 -0.01884355 0.3914175 -5.812492
                P.Value      adj.P.Val         B        SE
cg08991643 9.760006e-10 0.0003715956 10.420984 0.003697693
cg00472710 2.888752e-09 0.0004796229  9.358721 0.006209534
cg14962509 3.779207e-09 0.0004796229  9.095929 0.005050326
cg22454769 8.781002e-09 0.0007487501  8.272017 0.010578468
cg13772815 1.055254e-08 0.0007487501  8.092559 0.005035318
cg00228891 1.640026e-08 0.0007487501  7.662200 0.005942785
```

and this returns the DMRs using the dmrcate approach

```
> head(getAssociation(methRes, rid="dmrcate"))
                      coord no.cpgs       minfdr      Stouffer    maxbetafc
82    chr7:27152583-27155548      24 8.357043e-48 2.748206e-10 -0.04663463
86    chr8:22131899-22133356      13 1.659131e-38 7.231405e-09 -0.05363144
47 chr2:233251770-233252706       5 5.564853e-17 1.230653e-06 -0.04099509
15  chr11:72532891-72533664       3 2.282606e-16 1.768572e-06 -0.03545599
55 chr3:194014481-194015171       7 1.019507e-23 2.002484e-06 -0.04019226
13       chr11:315908-316456       3 1.455116e-14 2.585977e-06 -0.03351170
    meanbetafc
82 -0.02194618
86 -0.02950353
47 -0.03381141
15 -0.03003590
55 -0.01983233
13 -0.03227916
```

We can obtain the results of CpGs that mapped to some genes with the function `getGeneVals`. This function has the arguments `gene` and `genecol` to pass the names of the genes to be selected and the column name of feature data containing gene names. Let us illustrate how to test mean differences in all CpGs mapped to ARMS2. We can see in the rowData of meth that gene names are in the column `UCSC_RefGene_Name`:

```
> getGeneVals(methRes, "ARMS2", genecol = "UCSC_RefGene_Name",
+               fNames = c("chromosome", "start"))
```

We can also visually inspect the DMP results obtained from *limma* (DiffMean). For the association *P*-values, we compute a Q-Q plot, to evaluate the inflation of the estimates, likewise SNP associations, see Section 4.5.5. If estimates across the methylome are outside the confidence bands of the Q-Q plot then the model requires further covariates (e.g. sex, age, batch or cell composition). Methods described in Chapter 6 to obtain surrogate variables with *SVA* can be used for adjusting the model. The Q-Q plot is obtained by the generic function **plot** on the results. The arguments **rid** and **type** in the plot select the results of a given method and the type of plot to produce, for instance, to obtain the Q-Q plot for DMP analyses we run

```
> plot(methRes, rid = "DiffMean", type = "qq")
```

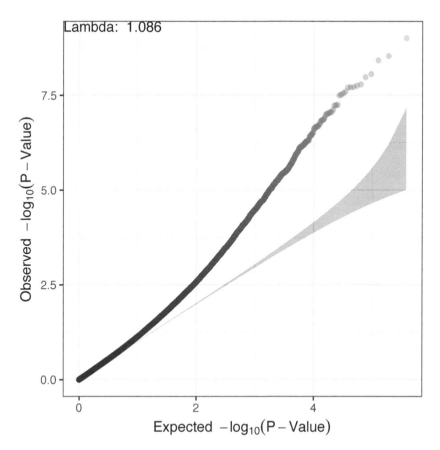

Including surrogate variables from a Surrogate Variable Analysis (SVA)indexsubjectindexSurrogate variable!analysis in the model can help to

correct the inflation (or deflation) of the methylome-wide estimates. The function runPipeline includes the parameter sva that calls sva from the *SVA* and automatically includes the surrogate variables as covariates in the linear model. Cell composition ratios are typically computed in surrogate variable analysis

```
> methRes <- runPipeline(set = gsm.grs.noNA,
+                          variable_names = "status",
+                          sva = TRUE)
```

Manhattan plots can also be obtained from a ResultSet object using the function plot and setting type to "manhattan". Manhattan plots are computed for DiffMean and DiffVar results. CpGs that belong to a region of interest can be highlighted through a GenomicRanges object in the highlight argument. For instance, we check the results of our Alzheimer's example in the *APOE* region whose genomic variability is associated with the disease.

```
> targetRange <- GRanges("chr19:45350000-45550000")
> plot(methRes, rid = "DiffMean", type = "manhattan",
+      main = "Differences in Means",
+      highlight = targetRange)


> plot(methRes, rid = "DiffVar", type = "manhattan",
+      main = "Differences in Variances",
+      highlight = targetRange)
```

Figures 8.1 and 8.2 show the regions with methylation differences in CpGs across the region and variance, respectively.

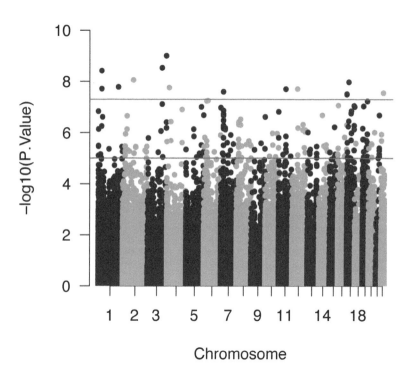

FIGURE 8.1
Manhattan plot corresponding to the comparison of mean CpG values between
cases and controls belonging to the Alzheimer's disease dataset (GEO number
GSE80970). Top horizontal lines stands for two different significant levels.

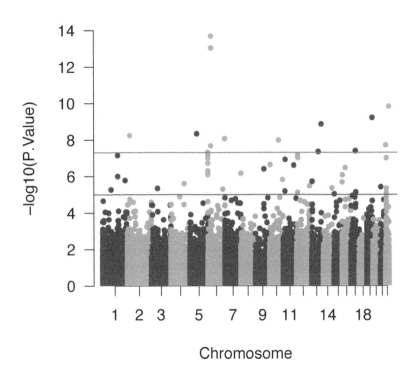

FIGURE 8.2
Manhattan plot corresponding to the comparison of variance CpG values between cases and controls belonging to the Alzheimer's disease dataset (GEO number GSE80970). Top horizontal lines stands for two different significant levels.

8.5   Methylation analysis of a target region

MEAL incorporates a method to explore a target region [129]. Redundancy analysis (RDA) is a multivariate method on the methylation status of the CpGs in a region of interest:

```
> targetRange <- GRanges("chr19:45350000-45550000")
> methRange <- runRDA(set = gsm.grs.noNA,
+                     model = ~ status,
+                     range = targetRange)
```

We can plot the RDA components with the function `plotRDA`, and color the samples by subject case-control status in the **pheno** argument

```
> plotRDA(object = methRange,
+         pheno = as.factor(gsm.grs.noNA$status))
```

Figure 8.3 shows similar distributions of normal and disease samples in the first RDA component. The model explains little variability $R^2 = 0.005$. More statistics of the RDA analysis can be obtained by:

```
> getRDAresults(methRange)
         R2        pval   global.R2 global.pval
0.005231010 0.138986101 0.004786719 0.274172583
```

`R2` and `pval` are the same values that are printed in the RDA plot. `global.R2` is the genome-wide variance explained by the model. `global.pval` is the probability that RDA components of a random region in the genome explain more variability than in the target region.

We can also plot the results obtained by other methods (DiffMean, Diff-Var, bumphunter, blockFinder and DMRcate) in the region of interest, using **plotRegion**. The plotting function requires a genomic range (e.g. object of class **GenomicRanges**) to be passed through the argument **range**:

```
> targetRange <- GRanges("chr19:45350000-45550000")
> plotRegion(rset = methRes, range = targetRange)
```

At the top of the Figure 8.4, we see the transcripts and CpGs annotations. The results of DMR detection methods can then be shown. At the bottom, we can find the results of two DMP analyses (DiffMean and DiffVar) (green). For our example, in particular, we do not see significant changes in methylation in any of the results.

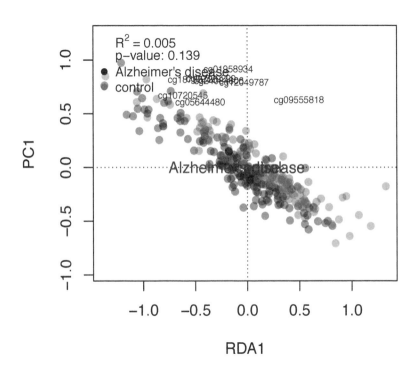

FIGURE 8.3
Methylation regional analysis (chr19:45.35-45.55 Mb) of Alzheimer case/control analysis (GEO number GSE80970)

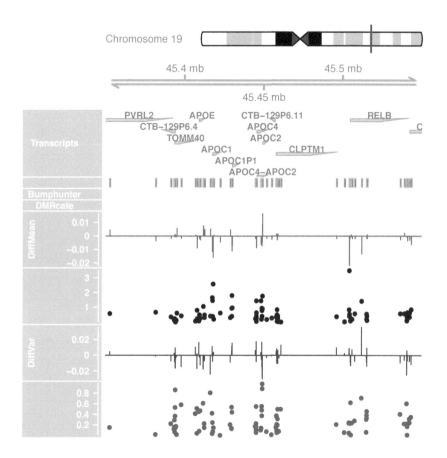

FIGURE 8.4
DMR analysis (top track after gene annotation) and DMP analysis for means
(DiffMean) and variance (DiffVar) of Alzheimer case/control analysis (GEO
number GSE80970). The coefficients (lines) and the *P*-values (dots) are also
depicted.

8.6 Epigenomic and transcriptomic visualization results

MEAL package can also perform transcriptomic analysis using the `runPipeline` function. The only difference is that an `ExpressionSet` is used as the input instead of a `GenomicRatioSet` object. This allows the user to encapsulate all the analyses described in Chapter 7 in a single function. It is recommended to change the annotation obtained from GEO and to the columns `start`, `end` and `chromosome`, as this information is used to create regional plots or to perform combined analyses of transcriptome and epigenome data.

We illustrate the use of *MEAL* to analyze the Alzheimer's transcriptomic study (GSE63061) of Chapter 7.

We load and format the data

```
> mask <- fData(gsm.expr)$Chromosome%in%c(1:22)
> gsm.expr <- gsm.expr[mask,]
>
> chr <- fData(gsm.expr)$Chromosome
> fData(gsm.expr)$Chromosome <- paste0("chr", chr)
>
> chrPos <- as.character(fData(gsm.expr)$Probe_Coordinates)
> ff <- function(x, pos){
+     ifelse(length(x)>1, x[pos], 1e10)
+ }
>
> temp <- sapply(strsplit(chrPos, "-"), FUN=ff, pos=1)
> chrStart <- sapply(strsplit(temp, ":"), "[[", 1)
>
> temp <- sapply(strsplit(chrPos, "-"), FUN=ff, pos=2)
> chrEnd <- sapply(strsplit(temp, ":"), "[[", 1)
> fData(gsm.expr)$start <- as.numeric(chrStart)
> fData(gsm.expr)$end <- as.numeric(chrEnd)
```

Once data is correctly formatted then we use `runPipeline` to perform the analyses

```
> status <- pData(gsm.expr)$characteristics_ch1
> status <- gsub("status: ","", as.character(status))
> pData(gsm.expr)$status <- relevel(as.factor(status),3)
> table(gsm.expr$status)

          CTL            AD borderline MCI       CTL to AD           MCI
          134            139              3               1           109
   MCI to CTL          OTHER
            1              1
> exprRes <- runPipeline(gsm.expr, variable_names = "status",
+                        sva=TRUE)
```

The function identifies the transcriptomic nature of the data and performs a transcriptome-wide association analysis. We create a volcano for the

`ResultSet` object, using the generic `plot` function. The features labels are automatically shown if *wordcloud* package is installed

```
> library(ggplot2)
> plot(exprRes, rid = "DiffMean", type = "volcano") +
+    ggtitle("Differences in Means")
```

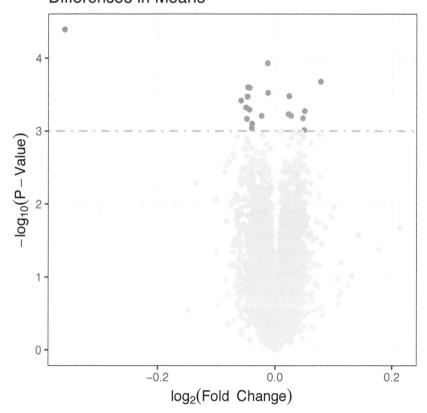

Results of multiple *omic* analyses can be visualized for a given region of interest toghether with genomic annotation. Figure 8.5 shows the results of the DMP and and DMR analysis, using differences in means (DiffMean) and bumphunter methods for epigenomic data and differentially expressed genes for transcriptomic data. The figure is obtained by:

```
> targetRange <- GRanges("chr19:45350000-45550000")
> plotRegion(rset = methRes, rset2 = exprRes,
+             range = targetRange,
+             results = c("DiffMean", "bumphunter"),
+             fNames = c("chromosome", "start", "end"),
+             fNames2 = c("Chromosome", "start", "end"))
```

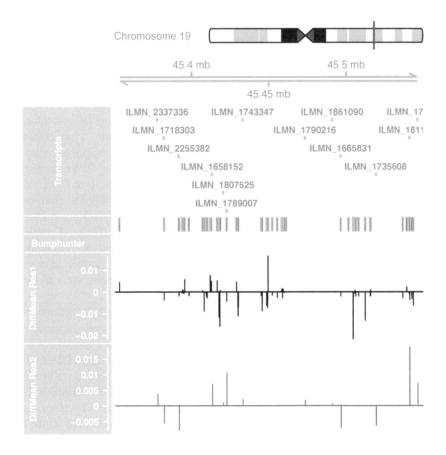

FIGURE 8.5
Epigenomic (DMRs and DMPs) and transcriptomic results obtained from Alzheimer's disease.

8.7   Cell proportion estimation

Methylation patterns strongly differ between tissues and cell types. As a result, methylomic data that is collected from complex tissues (e.g., blood, cord blood, saliva, etc.) contains a mixture of methylation signals from different cell types. The signals are weighted according to the mixing proportions of the individual's cell-type composition. When methlylomic data is obtained from heterogeneous sources for different individuals, methylome-wide association analyses may then reflect the differences between the sources rather than phenotypic differences. This is a situation that is analogous to the one described in Chapter 6 where cases and controls belong to different batches.

Accounting for cell-type composition can be addressed by two different approaches, which depend on whether a reference panel is considered or not. Methods that do not consider reference data include: 1) `RefFreeEWAS` [58] and SVA that are based on estimating unobserved variables that capture not only batch effects but also cell-type composition; 2) `Refactor` that uses principal component analysis on the 500 most informative CpG sites; 3) `RefFreeCellMix`, implemented in the R package *RefFreeEWAS*, that decomposes the total methylation sites into CpG-specific methylation states for a pre-specified number of cell types and subject-specific cell-type distributions [56]; and 4) `FaST-LMM-EWASher` that optimizes a spectral decomposition to estimate cell types [181]. All these methods provide a number of variables (e.g. principal components, surrogate variables, eigenvectors, etc.) to be used as covariates in association analyses.

The second approach includes methods that are based on reference panels. Houseman et al. [57] developed a method for cell-type correction that capitalizes on the idea that a panel of DMR can serve as a signature for the distribution of different types of white blood cells. It uses these DMRs as surrogates in a regression calibration to identify the cell mixture distribution. As regression calibration can lead to bias estimates, external validation data is used to calibrate the model. The validation dataset proposed by Houseman et al. consists of DNA-methylation data of sorted cell types from 46 white blood cell samples. There are other panels that can be used for other tissues. These are implemented in the package *meffil*[97] that can be installed from GitHub:

```
> library(devtools)
> install_github("perishky/meffil")
```

The function **meffil.estimate.cell.counts.from.betas** estimates the cell counts using methylation data (betas) and one of the following reference panels

```
> library(meffil)
> meffil.list.cell.type.references()
[1] "andrews and bakulski cord blood" "blood gse35069"
```

```
[3] "blood gse35069 chen"              "blood gse35069 complete"
[5] "cord blood gse68456"              "gervin and lyle cord blood"
[7] "saliva gse48472"
```

We illustrate how to estimate the cell counts in our example of Alzheimer's disease:

```
> betas <- getBeta(gsm.grs.noNA)
> cell.counts <- meffil.estimate.cell.counts.from.betas(betas,
+                          "blood gse35069 complete")
> head(cell.counts)
                Bcell      CD4T CD8T        Eos      Mono        Neu
GSM2139163 0.2888622 0.2474395    0 0.1252548 0.2658675 0.11733489
GSM2139164 0.2871643 0.2503917    0 0.1407267 0.2630283 0.09594465
GSM2139165 0.2707608 0.2510979    0 0.1188148 0.2772407 0.11002623
GSM2139166 0.2948644 0.2692381    0 0.1411104 0.2481792 0.10427831
GSM2139167 0.3023247 0.2639969    0 0.1330855 0.2561031 0.10730029
GSM2139168 0.2820867 0.2846518    0 0.1069999 0.2461567 0.13181548
                  NK
GSM2139163 0.1337515
GSM2139164 0.1430513
GSM2139165 0.1496122
GSM2139166 0.1238850
GSM2139167 0.1118446
GSM2139168 0.1286729
```

These variables can then be included as covariates in the association analysis and thus adjust for cell-type differences between individuals. We use the **runDiffMeanAnalysis** function from the *MEAL* package to illustrate how to perform an association analysis that adjusts for cell-type differences. Notice that this function has a different structure than **runPipeline** but it is faster since it only fits DMP analysis. Before running the analysis, we add cell-type estimates to the **GenomicRatioSet** object

```
> pData(gsm.grs.noNA) <- cbind(pData(gsm.grs.noNA),
+                         cell.counts) #add cell count variables
> model.cell <- ~ status + Bcell + CD4T + Eos + Mono +
+                 Neu + NK
> methRes.cell <- runDiffMeanAnalysis(gsm.grs.noNA,
+                                 model = model.cell)
```

Notice that this model adjusts only for cell-type composition. Extra variability associated with batch effects can be considered by setting the parameter **SVA=TRUE**. The top DMPs are

```
> head(getAssociation(methRes.cell, rid="DiffMean"))
                logFC         CI.L         CI.R   AveExpr         t
cg26600753 -0.02875156 -0.03861242 -0.01889070 0.8251298 -5.739493
cg00472710 -0.03712505 -0.04998820 -0.02426190 0.2457141 -5.681286
cg24366168  0.03005455  0.01945227  0.04065683 0.5953400  5.580050
cg21766308 -0.03831561 -0.05214953 -0.02448169 0.2078519 -5.452022
cg06221946 -0.01656554 -0.02255520 -0.01057587 0.4686716 -5.444148
```

```
cg08991643 -0.03405954 -0.04639645 -0.02172263 0.3141154 -5.434497
                  P.Value    adj.P.Val          B          SE
cg26600753 2.462365e-08 0.006367741 7.231487 0.003751917
cg00472710 3.344990e-08 0.006367741 6.932840 0.006043655
cg24366168 5.667463e-08 0.007192634 6.419196 0.004554837
cg21766308 1.092903e-07 0.007580574 5.780226 0.010738938
cg06221946 1.137522e-07 0.007580574 5.741318 0.004506213
cg08991643 1.194628e-07 0.007580574 5.693694 0.005425978
```

9

Exposomic studies

CONTENTS

9.1 Chapter overview

In this chapter, we study exposomic data. This type of data is different from the *omic* data that we discussed in the previous chapters. To start, the exposure variables that constitute the dataset are measured with different experimental techniques. They are highly diverse and heterogeneous, as they may include biological assays, obtained from personal questionnaires or derived from inferred exposures at geolocalized sites. Nevertheless, organisms respond to environmental changes and their efficiency in the response may result in phenotypic differences that may lead to adaptation or disease. The identification of the environmental changes is therefore essential to identify phenotypic differences among individuals and an unbiased scan on a large number of exposures is a starting point. In addition, exposure data may affect other molecular processes such as gene transcription or methylation. And therefore, important mechanisms may be discovered from the association between exposomic and transcriptomic, or methylomic data. Integration of *omic* data is then required. From the analysis perspective, exposomic data can be treated as the previous *omic* data, in which data structures are defined and methods are implemented on them. Here, we illustrate the use of *rexposome*, a comprehensive Bioconductor package with functions to impute, normalize and characterize correlational structures in exposomic data, to perform exposome-wise association analysis (ExWAS) and to test the association between exposomic and other *omic* variables.

9.2   The exposome

In addition to genetic risk factors, environmental exposures are essential contributors to the etiology of most complex diseases [35]. While individual genetic variants have been associated with complex traits, through the analysis of genomic data [55, 33, 26], a similar approach has been recently formulated for environmental factors; namely the analysis of exposomic data [170].

The exposome is loosely defined as the set of every exposure to which an individual is subjected from conception to death. While the exposome is incommensurable and it is not clear whether a specific biological state can be assigned to it, exposomic data, as a high-dimensional matrix of environmental exposures, is taken as a source to identify the environmental factors that may explain trait differences between individuals. Collection of high-dimensional exposure data requires consideration of both the nature of the exposures and their changes over time [170]. In particular, the nature of the exposures may be classified into three domains (Figure 9.1): internal environment, specific external environment and general external environment [171].

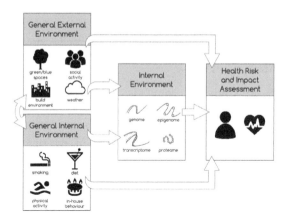

FIGURE 9.1
The effects and interactions between the different domains constituting the exposome and health risk.

A more concrete approach to an unbiased scan of exposures, may be given by the restriction to measuring the circulating chemicals in the body, which can also reflect both exogenous and endogenous exposures[118].

The study of the underlying mechanisms that link exposomic data with human health is an emerging research field with a strong potential to provide new insights into disease etiology [166]. Of particular interest is to explore how multiple exposures interact with multiple biological entities at different layers (e.g. transcriptome, methylome, proteome, metabolome, etc.) to increase our understanding in the development of complex traits [150].

9.2.1 Exposomic data

The **rexposome project**[1] provides a freely available framework for robust, scalable, reproducible and open-source development of methods to analyze exposomic data. The project offers R/Bioconductor packages for: (i) detailed characterization of the exposomic data, (ii) exposure signature discovery, (iii) linear and non-linear exposure-disease association studies and (iv) molecular *omic*-exposomic association and integration.

Multiomic data, including exposome, belonging to the INMA-Sabadell birth cohort that aims to study respiratory and neurodevelopmental disorders in children [47] is distributed with the *brgedata* package. It includes exposome `brge_expo`, transcriptome `brge_gexp`, methylome `brge_methy` and proteome `brge_prot` data

```
> library(brgedata)
> ls("package:brgedata")
 [1] "asthma"          "breastMulti"     "breastMulti_list"
 [4] "brge_expo"       "brge_gexp"       "brge_methy"
 [7] "brge_prot"       "genesAD"         "gwascatalog"
[10] "lusc"
```

that can be loaded into R using the **data** function. Exposome data can be loaded from text files as described in Section 3.7. The object `brge_expo` is an *ExposomeSet* that has been created in the same way using data from the INMA cohort:

```
> brge_expo
Object of class 'ExposomeSet' (storageMode: environment)
 . exposures description:
    . categorical:  0
    . continuous:  15
 . exposures transformation:
    . categorical: 0
    . transformed: 0
    . standardized: 0
    . imputed: 0
 . assayData: 15 exposures 110 individuals
    . element names: exp, raw
    . exposures: Ben_p, ..., PCB153
    . individuals: x0001, ..., x0119
 . phenoData: 110 individuals 6 phenotypes
    . individuals: x0001, ..., x0119
    . phenotypes: Asthma, ..., Age
 . featureData: 15 exposures 12 explanations
    . exposures: Ben_p, ..., PCB153
    . descriptions: Family, ..., .imp
```

---

[1]https://isglobal-brge.github.io/rexposome/

```
experimentData: use 'experimentData(object)'
Annotation:
```

The object **brge_expo** is of *ExposomeSet* class (see Section 3.7 in Chapter 3), it contains information about 110 samples and 15 exposome variables grouped into 3 families of exposures (air pollution, BPA and PCBs). The transcriptomic data (**brge_expo**) is for 75 individuals using the HTA 2.0 array (Affymetrix, USA), which provides gene expression for 67,528 transcript clusters. The methylome brgedata (**brge_methy**) consists of 476,946 features obtained with the 450k methylation array (Illumina, USA) and measured in 115 individuals. Forty-seven proteins (**brge_prot**) were measured in 90 individuals that are also available from *brgedata*. Each *omic* data contains a different number of samples.

9.3   Exposome characterization

The package *rexposome* has a number of methods to characterize exposomic data

```
> library(rexposome)
```

The functions **tableMissings** and **plotMissings** can be used to deal with missing information on the exposomic and phenotypic data within the *ExposomeSet* structure

```
> tableMissings(brge_expo, set = "phenotypes")
    Sex   Asthma Rhinitis     Age Whistles      CBMI
      3        7        7       7        8         9
```

For this data there are missing values in the phenotype data. Missing data in the exposures is visualized in Figure 9.2, obtained by the *ggplot2* package.

```
> library(ggplot2)
> plotMissings(brge_expo, set = "exposures") +
+     ggtitle("Missing Values in INMA-Sabadell Exposome")
```

We observe that, in general, exposures have less than 25% of missing data.

Imputation data may be required for analyses that require complete cases. We use the packages *mice* and *Hmisc* for imputation [14], that are called within the **imputation** function of *rexposome*. **imputation** is designed to impute missing values in exposures, not in phenotypes

```
> brge_expo_complete <- imputation(brge_expo)
> tableMissings(brge_expo_complete, set="exposures")
 Ben_p Ben_t1 Ben_t2 Ben_t3  BPA_p BPA_t1 BPA_t3   NO2_p NO2_t1 NO2_t2
```

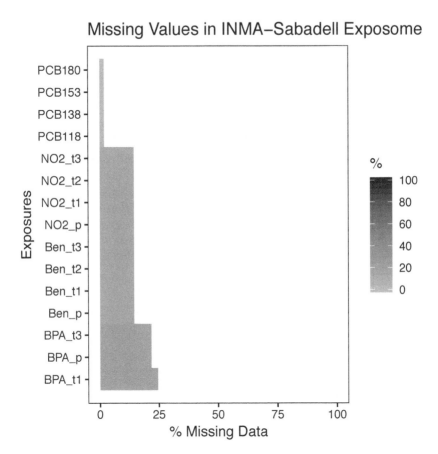

FIGURE 9.2
Missing data pattern of the *brgedata* exposome dataset.

```
    0       0       0       0       0       0       0       0       0       0
NO2_t3 PCB118 PCB138 PCB153 PCB180
    0       0       0       0       0
```

Imputation quality is sensitive to the data structure and scientific question. By default, imputation is performed with *Hmisc*. Multiple imputations from *mice* can also be used; see *mice*'s vignette[2].

After imputation, we can visualize the distribution of the data using `plotFamily` (Figure 9.3).

```
> plotFamily(brge_expo, family = "all")
```

A similar figure can be obtained stratifying by any categorical variable. For instance, Figure 9.4 shows exposure distribution of PCBs stratified by sex.

```
> plotFamily(brge_expo, family = "PCBs", group = "Sex")
```

It can be seen that the range of different exposures may vary. *rexposome* has implemented several methods of normalization of continuous exposures including normal, robust or interquartile range (IQR) normalizations, among others. Normalization across exposomic data is important for comparing the effects of different exposures on a phenotype or to perform multivariate analyses. The normal method standardizes the exposures by scaling and homogenizing by the mean and standard deviation. Robust normalization uses the median and the median absolute deviation, while IQR is based on the median and the inter-quantile range:

```
> brge_expo_standard <- standardize(brge_expo)
> brge_expo_standard
Object of class 'ExposomeSet' (storageMode: environment)
 . exposures description:
   . categorical:  0
   . continuous:  15
 . exposures transformation:
   . categorical: 0
   . transformed: 0
   . standardized: 15
   . imputed: 0
 . assayData: 15 exposures 110 individuals
   . element names: exp
   . exposures: Ben_p, ..., PCB153
   . individuals: x0001, ..., x0119
 . phenoData: 110 individuals 6 phenotypes
   . individuals: x0001, ..., x0119
   . phenotypes: Asthma, ..., Age
 . featureData: 15 exposures 12 explanations
   . exposures: Ben_p, ..., PCB153
   . descriptions: Family, ..., .imp
experimentData: use 'experimentData(object)'
Annotation:
```

---

[2]https://bioconductor.org/packages/3.7/bioc/vignettes/rexposome/inst/doc/mu tiple_imputation_data_analysis.html

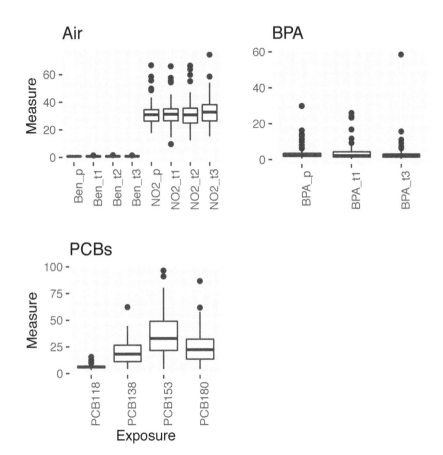

FIGURE 9.3
Distribution of exposures accross families on brge exposome dataset.

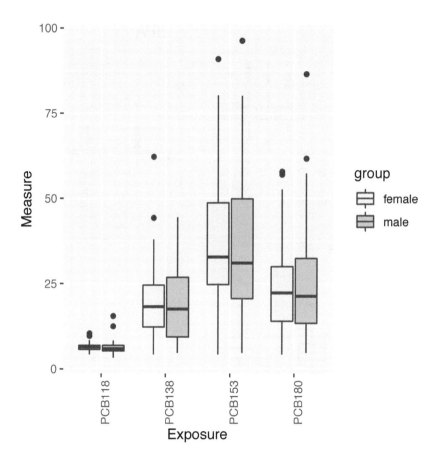

FIGURE 9.4
Distribution of PCBs stratified by sex the *brgedata* exposome dataset.

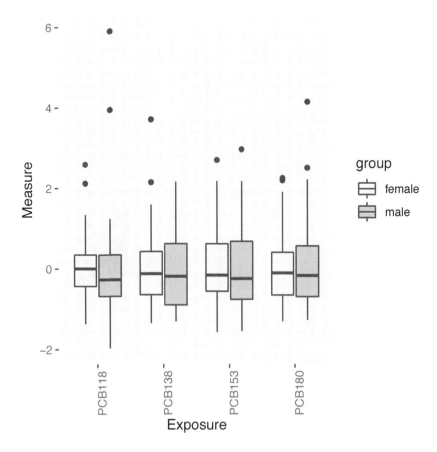

FIGURE 9.5
Distribution of PCBs estratified by sex on the *brgedata* exposome dataset.

A figure describing standardized data can also be obtained using the `plotFamily` function using the object `brge_expo_standard` as input. Figure 9.5 shows standardized PCB exposures by sex.

Once exposure data have been standardized, a principal component analysis (PCA) can be performed to determine whether exposures may be clustered in the two first components

```
> brge_expo_pca <- pca(brge_expo_standard)
> rexposome::plotPCA(brge_expo_pca, set = "all")
```

The top-left panel of Figure 9.6 shows a clustering of the exposures in the first two eigen-subjects, which explain 60% of total variability (bottom panels). The top-right panel shows the projection of individuals into first principal components (eigen-variables). This figure can be colored by a categorical

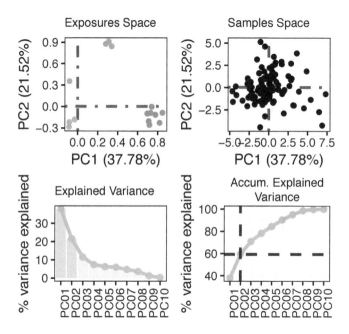

FIGURE 9.6
Principal component analysis on the *brgedata* exposome standardized data

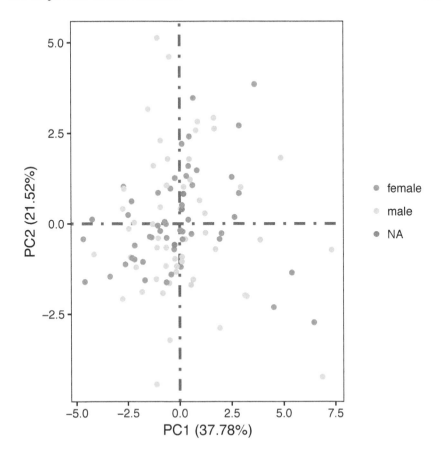

FIGURE 9.7
Principal component analysis on the *brgedata* exposome standardized data using gender as an illustrative variable.

variable that may represent a phenotype (Figure 9.7). Notice that we write `rexposome::plotPCA` since the function `plotPCA` is also available from other R packages:

```
> rexposome::plotPCA(brge_expo_pca,
+                    set = "samples",
+                    phenotype = "Sex")
```

An important feature across the expoposomic data is its correlational structure, which can be obtained with the `plotCorrelation`. Figure 9.8 shows the correlation matrix among exposures grouped by families. Notice that a circos plot could also be computed by setting `type = "circos"`.

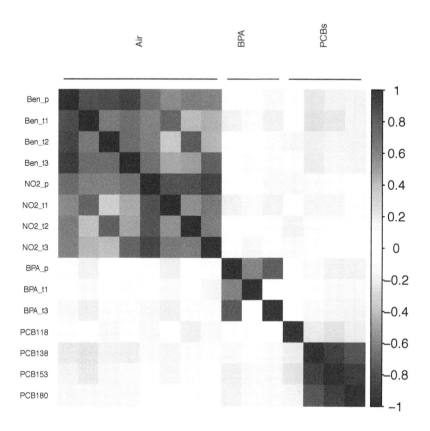

FIGURE 9.8
Correlation matrix between the *brgedata* exposure datasets.

```
> exp_cr <- correlation(brge_expo_standard,
+                       use = "pairwise.complete.obs",
+                       method.cor = "pearson")
> plotCorrelation(exp_cr, type = "matrix")
```

9.4   Exposome-wide association analyses

Our initial interest is to find whether there is any exposure in the exposomic data that significantly associates with a phenotype. The exposome-wide association analysis (ExWAS) is therefore equivalent to the other *omic* association analyses, such as GWAS. The `exwas` function performs the massive univariate testing, where a generalized linear model is fit for each exposure [106]. The function's syntax is similar to the basic R function `glm`. The function allows the use of splines as covariates, using the fucntion `bs` from the *splines* package, when non-linearity is expected. We then perform ExWASa analysis on our examples, where the outcome of interest is asthma and we adjust by sex and age

```
> library(splines)
> resExwas <- exwas(brge_expo_standard,
+                   formula = Asthma ~ Sex + Age,
+                   family = "binomial")
> resExwas
An object of class 'ExWAS'

        Asthma ~ Sex+Age

Tested exposures:  15
Threshold for effective tests (TEF):  6.97e-03
  . Tests < TEF: 0
```

The result of the analysis in the `resExwas` object output shows the number of exposures that pass the significance level, corrected for multiple testing [82]. In our case, no significant results are obtained. The *P*-values of association can be plotted in a Manhattan plot of exposures (Figure 9.9)

```
> clr <- rainbow(length(familyNames(brge_expo_standard)))
> names(clr) <- familyNames(brge_expo_standard)
> plotExwas(resExwas, color = clr) +
+   ggtitle("ExWAS - Univariate Approach")
```

The exposure effects with their confidence intervals can also be plotted with the function `plotEffec` on the results, as in Figure (9.10)

```
> plotEffect(resExwas)
```

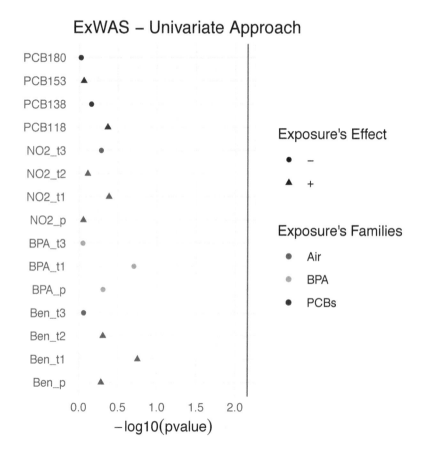

FIGURE 9.9
Manhattan plot of exposures obtained from ExWAS.

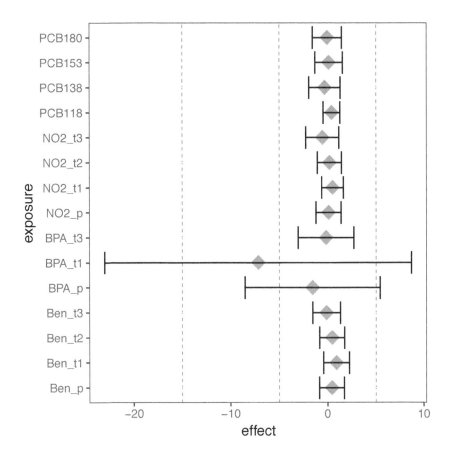

FIGURE 9.10
Effects (beta values) and confidence intervals of exposures corresponding to the asthma EXWAS example.

9.5   Association between exposomic and other *omic* data

As environmental exposures are essential factors for trait differences between individuals, they are also expected to explain individual differences at lower biological levels. As such, environmental exposures may be associated with the development of a phenotype by affecting, for instance, the transcription or methylation status of genes. In such cases, one is then interested in investigating the extent to which exposomic differences can explain transcriptomic differences between individuals. The package *omicRexposome* includes methods to analyze multiple *omic* dataset structures in an `MultiDataSet` class, in which exposomic data is encoded in a `ExposomeSet`, transcriptomic data in a `ExpressionSet`, methylomic data in a `MethylationSet` or `GenomicRatioSet`.

Massive association analysis tests each exposure independently against all the features available in the *omic* tables using the methods of the *limma* package, described in Chapter 7. The univariate testing is obtained from linear regression models where exposures are continuous and the coefficients of the regression models are interpreted as changes in *omic* features by one unit change in the exposure.

9.5.1 Exposome-transcriptome data analysis

Within the `association` function, one can perform an exposome-transcriptome analysis on a *MultiDataSet*, where exposomic data is encoded in an *ExposomeSet* and transcriptomic data in an *ExpressionSet*. The function requires the names of the exposome and gene expression datasets, and the formula indicating covariates. Analyses are performed for common samples between datasets. For fitting the models

$$transcript_i = \beta_1 exposure_j + \beta_2 Sex + \beta_3 Age + \sum_{k=1}^{K} \beta_{3+k} SV_k$$

where $i = 1, \ldots, 67528$ are the transcripts, $j = 1, \ldots, 15$ are the different exposures and K are the surrogate variables (SV) estimated using the fast SVA method and. The different associations are estimated by executing:

```
> library(omicRexposome)
> library(MultiDataSet)
> mds <- createMultiDataSet()
> mds <- add_exp(mds, brge_expo_standard)
> mds <- add_genexp(mds, brge_gexp)
> resGene <- association(mds, formula = ~ Sex + Age,
+                        sva = "fast",
+                        expset = "exposures",
+                        omicset = "expression")
> resGene
Object of class 'ResultSet'
 . created with: association
   . via: expression and exposures
 . sva:  NA
 . #results: 15 ( error: 0 )
 . featureData:  2
   . expression: 67528x11
   . exposures: 15x12
```

From the results object **resGene**, we can retrieve the transcripts that pass a given significance threshold

```
> tableHits(resGene, th = 1e-06)
        exposure hits
Ben_p      Ben_p   0
Ben_t1    Ben_t1   0
Ben_t2    Ben_t2   0
Ben_t3    Ben_t3   0
BPA_p      BPA_p   0
BPA_t1    BPA_t1   1
BPA_t3    BPA_t3   0
NO2_p      NO2_p   0
NO2_t1    NO2_t1   0
NO2_t2    NO2_t2   0
NO2_t3    NO2_t3   0
PCB118    PCB118   0
PCB138    PCB138   0
```

```
PCB153    PCB153    0
PCB180    PCB180    0
```

For each analysis the `association` function computes the Q-Q plot parameter λ, to determine whether the estimates are inflated. λ is estimated likewise GWASs [21], using a trimmed robust regression defined by the argument `trim`, which controls the percentage of P-values used in the robust regression [164].

```
> tableLambda(resGene, trim = 0.95)
          exposure    lambda
Ben_p      Ben_p 1.0014232
Ben_t1    Ben_t1 1.0030942
Ben_t2    Ben_t2 1.0057553
Ben_t3    Ben_t3 1.0008209
BPA_p      BPA_p 0.9716879
BPA_t1    BPA_t1 0.9506279
BPA_t3    BPA_t3 0.9898204
NO2_p      NO2_p 1.0045770
NO2_t1    NO2_t1 1.0018950
NO2_t2    NO2_t2 1.0020010
NO2_t3    NO2_t3 0.9759471
PCB118    PCB118 1.0550025
PCB138    PCB138 1.0109964
PCB153    PCB153 0.9838635
PCB180    PCB180 0.9851821
```

Both volcano (Figure 9.11) and Q-Q plots (Figure 9.12) can be drawn using the `plotAssociation` function. This is the code for the volcano plot

```
> plotAssociation(resGene, rid = "NO2_p", type = "volcano",
+                 tPV = -log10(1e-04), tFC = 0.1,
+                 show.effect=FALSE) +
+    ggplot2::ggtitle("NO2_p")
```

and this for the Q-Q plot

```
> plotAssociation(resGene, rid = "NO2_p", type = "qq",
+                 show.lambda=FALSE) +  ggplot2::ggtitle("NO2_p")
```

An alternative strategy of exposome-transcriptome analysis is first to reduce the dimensionality of the exposome data in a subject-wise clustering, and then regress the clustering status on the transcriptome. The aim is then identifying any gene transcription that significantly associates with the given summarization of the exposomic data. There are numerous clustering methods in R (see `?clustering`) such as those implemented in the *hclust*, *mclust* or *flexmix* packages. In our example, we use *mclust* that is indicated by `method=Mclust`. Notice that this method requires complete cases

```
> brge_expo_complete <- imputation(brge_expo)
> brge_expo_standard2 <- standardize(brge_expo_complete)
```

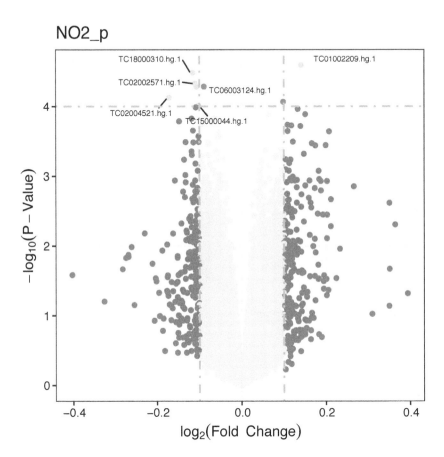

FIGURE 9.11
Volcano of NO2 and transcriptomic data from *brgedata* exposome data example.

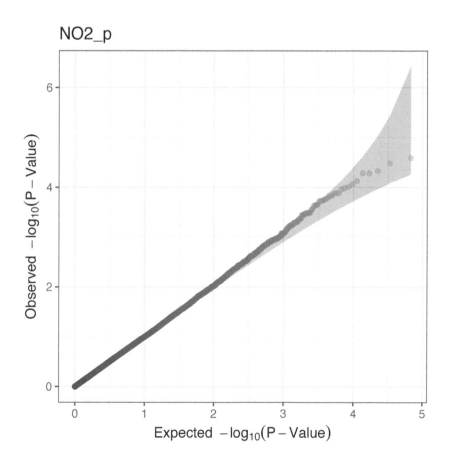

FIGURE 9.12
QQ plot of NO2 and transcriptomic data from *brgedata* exposome data example.

```
>
> brge_expo_cluster <- clustering(brge_expo_standard2,
+                                        method = Mclust,
+                                        G=2)
> table(rexposome::classification(brge_expo_cluster))

  1   2
 93  17
```

Figure 9.13 shows the clustering

```
> plotClassification(brge_expo_cluster)
```

We see that the clustering corresponds to a group highly exposed to benzene, and little to BPAs and NO2, and a second group with the opposite pattern. We then analyze the exposome-transcriptome association taking the group clustering as a predictor of transcriptomic differences

```
> mds2 <- createMultiDataSet()
> mds2 <- add_cls(mds2, brge_expo_cluster)
> mds2 <- add_genexp(mds2, brge_gexp)
> resGene2 <- association(mds2, formula = ~ Sex + Age,
+                           sva = "fast",
+                           expset = "cluster",
+                           omicset = "expression")
> tableHits(resGene2, th = 1e-06)
         exposure hits
cluster  cluster    0
```

We can also plot the results

```
> par(mfrow=c(2,1))
> plotAssociation(resGene2, type = "volcano",
+                   tPV = -log10(1e-06), tFC = 1,
+                   show.effect=FALSE) +
+    ggplot2::ggtitle("Transcriptome - Cluster group 1 vs 2")


>
> plotAssociation(resGene2, type = "qq", show.lambda=FALSE) +
+    ggplot2::ggtitle("NO2_p")
```

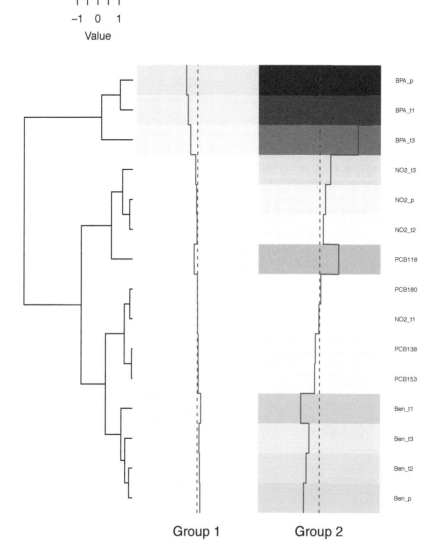

FIGURE 9.13
Exposome clustering using the mclust method including exposures from the
brgedata example.

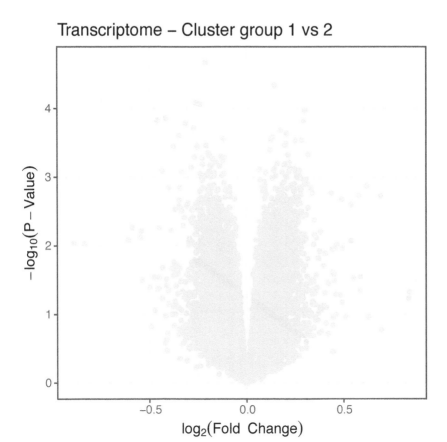

FIGURE 9.14
Volcano and QQ plots of transcriptomic data analysis considering exposome
clusters as the factor variable.

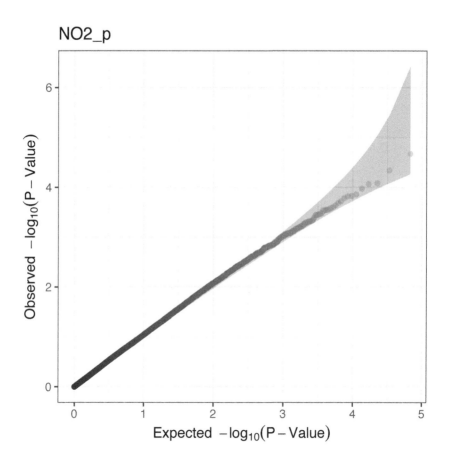

FIGURE 9.15
Volcano and QQ plots of transcriptomic data analysis considering exposome clusters as the factor variable.

9.5.2 Exposome-methylome data analysis

Association between the exposome and methylome follows the same pipeline previously described. We can add to the existing *MultiDataSet* objects, the methylation data and then perform the analyses using the same procedure:

```
> mds <- add_methy(mds, brge_methy)
> resMethy <- association(mds, formula = ~ Sex + Age,
+                         sva = "fast",
+                         expset = "exposures",
+                         omicset = "methylation")
> mds2 <- add_methy(mds2, brge_methy)
> resMethy2 <- association(mds2, formula = ~ Sex + Age,
+                          sva = "fast",
+                          expset = "cluster",
+                          omicset = "methylation")
```

We can produce specific volcano plots for a given exposure, for instance, for the association between NO2 and the methylome see Figure 9.16

```
> plotAssociation(resMethy, rid = "NO2_p", type = "volcano",
+                 tPV = -log10(1e-04), tFC = 0.1,
+                 show.effect=FALSE) +
+   ggplot2::ggtitle("NO2_p")
```

Figure 9.17 shows the results comparing the methylation levels between the exposure groups obtained after applying a clustering method to the whole exposome:

```
> plotAssociation(resMethy2, type = "volcano",
+                 tPV = -log10(1e-08), tFC = 0.3,
+                 show.effect=FALSE) +
+   ggplot2::ggtitle("Epigenome - Exposome cluster group 1 vs 2")
```

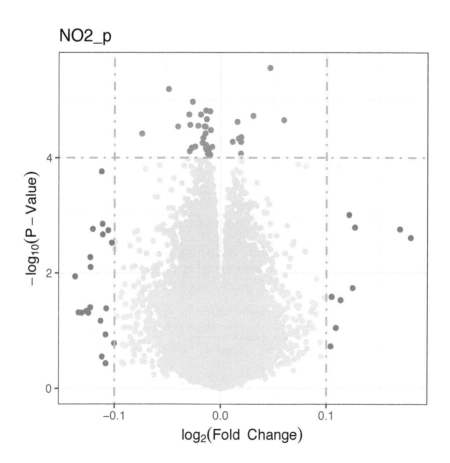

FIGURE 9.16
Volcano plot of NO2 analysis on epigenomic data.

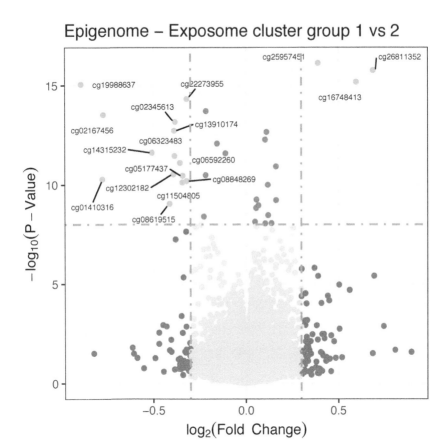

FIGURE 9.17
Volcano plot of epigenomic data comparing exposome clusters.

10

Enrichment analysis

CONTENTS

10.1 Chapter overview

In this chapter, we demonstrate how to perform enrichment analyses of *omic*-wise association analysis. The identification of single variables in *omic* data through massive univariate testing is clearly a simplistic approach to study biological systems that are inherently complex. Its strength resides in the scope and the easiness of the approach to identify landmarks in the data. A frequent result from these studies is the identification of several variables that are significantly associated with the phenotype of interest. In addition, as more individuals are included in the studies, more significant associations are identified. How do we make sense of the observed associations? Enrichment analyses try to answer the question of whether the emerging pattern of associations can be mapped to known biological functions.

The are two predominantly used enrichment methods. One is known as over representation analysis (ORA) which tests whether a gene set contains disproportional many genes of significant expression change in a given gene set. The second method, called gene set enrichment analysis (GSEA), assesses whether genes of a gene set accumulate at the top or bottom of the full gene vector ordered by direction and magnitude of expression change [149]. However, the term gene set enrichment analysis nowadays subsumes a general strategy implemented by a wide range of methods [60]. Those methods have in common the same goal, although approach and statistical model can vary substantially [69]. In this chapter we will focus on ORA approach which is commonly used in practice since it can be extended to any gene set as those defined, for instance, at Molecular Siganture Database (MSigDb).

We illustrate the practical issues of the approach on the transcriptomic study of Alzheimer's disease, discussed in Chapter 7. We particularly demonstrate how to annotate the results of the association analysis to the reference genome, using Bioconductor's annotation databases (.db), how to map the results to biological functions curated in databases like GO and KEGG, and how to assess whether the mapping, or enrichment, is significant, using the Bioconductors package *GOstats* and *clusterProfiler*.

We also outline how to perform these type of analysis in the CNV settings where the outcome is a list of genomic ranges instead of genes. In such situations, one can be insterested in deciphering whether the resulting CNV regions overlap with functional genomic regions such as genes, promoters, or enhancers. Finally, we explain the use of *CTDquerier* to study whether the significant associations are enriched in diseases and interactions with chemicals.

10.2   Enrichment analysis and statistical power

The aim of genome, transcriptome and epigenome-wide association studies is the identification of *any* variable, or groups of variables, that is significantly associated with phenotypic differences among individuals. The studies are based on hypothesis-free analyses. That is, they are scientifically agnostic, in the sense that they do not specifically test a presumed state of affairs of a biological system, but rather they test how unusual each of the correlations between the trait and the *omic* variables really is. As such, *omic* association studies are used to generate scientific hypotheses and, in particular, the interpretation of their results are left for posterior inquiries.

Another important limitation of *omic* association studies relates to the assessment of statistically significant correlations of high-dimensional data. As we have mentioned before, the results need to be corrected for the number of tests performed. Take for instance a study that tries to identify whether there is *any* significant correlation of a trait with 100 variables. If there is no real association with any of such variables other than chance then, from a 100 tests, we can expect that 5 of them will be significant, at 5% significance level. The probability of finding at least one significant association, a false positive in this case, is close to 1 $(0.994 = 1 - (1 - 0.05)^{100})$. To reduce the chance of false positives down to 5%, the significance level of a single association should be divided by the number of tests. For Bonferroni correction, we drop the level to 5%/100 in our example, reducing the probability of observing at least one false positive to the desired 5% $(0.048 = 1 - (1 - 0.05/100)$. For high dimensional data of around one million variables the corrected level of significance should be scaled down to 5×10^{-8}. The required low threshold comes at the expense of greatly increasing the number of subjects to the order of hundreds of thousands when the expected associations are of 1% increase risk.

Enrichment analyses on association studies address the difficulties of interpretation and of statistical power. The strategy of enrichment analyses is to loosen the agnostic demand by considering groups of *omic* variables that are known to belong to the same biological functions. An enrichment analysis then performs a single test for the association between the trait and a subgroup of *omic* variables. The set of biological groups is much smaller than that of the *omic* variables and therefore the statistical threshold of significance for individual groups can be loosened in about three orders of magnitude.

10.3   Gene set annotations

Genes acting in different biological functions, cellular components and molecular pathways have been curated into gene ontologies (GO), and biochemical pathways that are accessible from the GO, KEGG (Kyoto encylopedia of genes and genomes) and Biocarta initiatives, among others. Based on this information, *omic* variables are mapped to the genes that have been grouped into the different sets. For instance, a group of transcriptomic variables may correspond to all the transcripts of the genes whose proteins interact in a carbohydrate metabolism, while a group of SNPs can be selected from the genes that translate into brain physiology. *Omic* variables are mapped to the closest genes at neighboring distances, as given by the genomic coordinates of a reference genome. Therefore, not all variables are mapped and, as genes belong to more than one gene-set, some variables will be selected more than once. Annotation packages for the *omic* variables, gene-pathways and numerous genomic features are accessible in Bioconductor and updated every six months against the latest reference genome.

The first step in an enrichment analysis is to map the *omic* variables to the functional pathways. We demonstrate how query annotation data from the human genome to retrieve, for instance, the GO terms associated to a gene or the genes that belong to a given GO term. The example will allow us to show how Bioconductor annotation packages are stored and used. The annotation package for the human genome is stored in Bioconductor's database library *org.Hs.eg.db*

```
> library(org.Hs.eg.db)
> org.Hs.eg.db
OrgDb object:
| DBSCHEMAVERSION: 2.1
| Db type: OrgDb
| Supporting package: AnnotationDbi
| DBSCHEMA: HUMAN_DB
| ORGANISM: Homo sapiens
```

```
| SPECIES: Human
| EGSOURCEDATE: 2018-Apr4
| EGSOURCENAME: Entrez Gene
| EGSOURCEURL: ftp://ftp.ncbi.nlm.nih.gov/gene/DATA
| CENTRALID: EG
| TAXID: 9606
| GOSOURCENAME: Gene Ontology
| GOSOURCEURL: ftp://ftp.geneontology.org/pub/go/godatabase/archive/latest-lite/
| GOSOURCEDATE: 2018-Mar28
| GOEGSOURCEDATE: 2018-Apr4
| GOEGSOURCENAME: Entrez Gene
| GOEGSOURCEURL: ftp://ftp.ncbi.nlm.nih.gov/gene/DATA
| KEGGSOURCENAME: KEGG GENOME
| KEGGSOURCEURL: ftp://ftp.genome.jp/pub/kegg/genomes
| KEGGSOURCEDATE: 2011-Mar15
| GPSOURCENAME: UCSC Genome Bioinformatics (Homo sapiens)
| GPSOURCEURL:
| GPSOURCEDATE: 2018-Mar26
| ENSOURCEDATE: 2017-Dec04
| ENSOURCENAME: Ensembl
| ENSOURCEURL: ftp://ftp.ensembl.org/pub/current_fasta
| UPSOURCENAME: Uniprot
| UPSOURCEURL: http://www.UniProt.org/
| UPSOURCEDATE: Mon Apr  9 20:58:54 2018
```

The library loads on the R session a number of datasets that can be listed

```
> datasets <- ls("package:org.Hs.eg.db")
> head(datasets)
[1] "org.Hs.eg"          "org.Hs.eg.db"       "org.Hs.eg_dbconn"
[4] "org.Hs.eg_dbfile"   "org.Hs.eg_dbInfo"   "org.Hs.eg_dbschema"
```

Specific information on the source and release of each annotation dataset can be obtained by consulting its help page. For instance, we can find the specific information relating the dataset org.Hs.egCHR

```
> help(org.Hs.egCHR)
```

Annotation packages for different organisms are offered in Bioconductor as AnnotationDb objects, whose datasets are stored, mapped and retrieved using a common set of methods. The list of accessible data is obtained in the object using the function columns

```
> cls <- columns(org.Hs.eg.db)
```

As the types of datasets are also documented, we find that the entry SYMBOL

```
> help("SYMBOL")
```

corresponds to the official gene symbol. Datasets coordinated by the AnnotationDb object are mapped and retrieved for a list of selected variables that correspond to the entries of a given keytype. For instance, we can find the keytypes than can be selected for mapping

```
> keytypes(org.Hs.eg.db)
 [1] "ACCNUM"        "ALIAS"        "ENSEMBL"      "ENSEMBLPROT"
 [5] "ENSEMBLTRANS" "ENTREZID"     "ENZYME"       "EVIDENCE"
 [9] "EVIDENCEALL"  "GENENAME"     "GO"           "GOALL"
[13] "IPI"          "MAP"          "OMIM"         "ONTOLOGY"
[17] "ONTOLOGYALL"  "PATH"         "PFAM"         "PMID"
[21] "PROSITE"      "REFSEQ"       "SYMBOL"       "UCSCKG"
[25] "UNIGENE"      "UNIPROT"
> k <- head(keys(org.Hs.eg.db, keytype="SYMBOL"))
> k
[1] "A1BG"  "A2M"   "A2MP1" "NAT1"  "NAT2"  "NATP"
```

The gene symbol is one the variables that can be queried across datasets. Using the function `keys` with argument `keytype="SYMBOL"`, we obtain the list of all the genes annotated in the database. With the function `select`, we retrieve all the information stored in `columns` for a list of keys. For instance, we can select the variables `MAP`, `ENTREZID` and `GO` for the gene *APP* which is a key of value `APP` of type `SYMBOL`.

```
> mapGenes <- select(org.Hs.eg.db, keys = "APP",
+                     columns=c("MAP","ENTREZID","GO"),
+                     keytype = "SYMBOL")
>
> head(mapGenes)
  SYMBOL      MAP ENTREZID         GO EVIDENCE ONTOLOGY
1    APP 21q21.3      351 GO:0001934      IDA       BP
2    APP 21q21.3      351 GO:0001934      IGI       BP
3    APP 21q21.3      351 GO:0001967      IEA       BP
4    APP 21q21.3      351 GO:0002265      IGI       BP
5    APP 21q21.3      351 GO:0002576      TAS       BP
6    APP 21q21.3      351 GO:0003677      ISS       MF
```

While we have retrieved data for the single key *APP*, vectors of keys will return data for each key, all concatenated in a single `data.frame`. In particular, we obtained the cytoband location of *APP* (`MAP`), the entrez gene identifier (`ENTREZID`) and the variable `GO` storing the gene ontologies associated with the gene.

The Gene Ontology Project (`http://www.geneontology.org`) provides a controlled vocabulary to describe gene and gene product attributes in any organism. This vocabulary consists of "GO terms, which are pairs of term identifier (GO ID) and description, for instance

- GO:0042594 -response to starvation

- GO:0003700 -transcription

- GO:0007154 -cell communication

Each GO term is assigned to one of the following three ontologies: Biological Process (BP), Molecular Function (MF) or Cellular Component (CC). A gene product might be associated with one or more cellular components as it is active in one or more biological processes, in which it performs diverse molecular functions. Full details are described at http://geneontology.org/

We can also retrieve from org.Hs.eg.db the symbols and entrez identifiers of the genes that belong to the GO term GO:0042594. Notice that we write AnnotationDbi::select since the function select is available from different R packages:

```
> genesGO <- AnnotationDbi::select(org.Hs.eg.db,
+                   keys="GO:0042594",
+                   columns=c("ENTREZID","SYMBOL"),
+                   keytype="GO")
> head(genesGO)
          GO EVIDENCE ONTOLOGY ENTREZID  SYMBOL
1 GO:0042594      IEA       BP       38    ACAT1
2 GO:0042594      IEA       BP      116  ADCYAP1
3 GO:0042594      IEA       BP      133      ADM
4 GO:0042594      IEA       BP      158     ADSL
5 GO:0042594      IEA       BP      790      CAD
6 GO:0042594      IEA       BP      867      CBL
```

An AnnotationDbi package GO.db is also available for GO annotations, where we can, for instance, retrieve the definition for a given GO term

```
> library(GO.db)
> def <- AnnotationDbi::select(GO.db,
+                             keys="GO:0042594",
+                             columns=c("DEFINITION"),
+                             keytype="GOID")
> head(def)
         GOID
1 GO:0042594

1 Any process that results in a change in state or activity of a cell
or an organism (in terms of movement, secretion, enzyme production,
gene expression, etc.) as a result of a starvation stimulus, deprivation
of nourishment.
```

Other possibilities to annotate genes available in Bioconductor are, for instance, TxDb objects such as *TxDb.Hsapiens.UCSC.hg19.knownGene* library [19] or annotation databases generated from Ensembl such as *EnsDb.Hsapiens.v86* package [116].

10.4   Over representation analysis

In general enrichment analysis of transcriptomic data, we want to identify the biochemical pathways of the genes that significantly changed their gene expression between two experimental conditions. In such analysis, different GO terms representing overlapping functions will be mapped by numerous common genes. Some GO terms have also hierarchical relations and therefore all the genes in one GO term may be contained in another. In addition, one gene can radically change its function depending on internal or external context. As such, the identification of the GO terms that are most likely affected by the significant changes in their *omic* variables needs to overcome a number of issues. While this is an area of active research, there are two main approaches to the problem. The first one is to compute the significance of a GO term and account for its neighboring GO relationships [1]. In this case, GO terms are endowed with self-contained statistics given by its gene elements. The second and the most popular approach does not depend on the specific genes in the GO term, nor in the inter-relations between them. A test statistic is given to any subset of genes that measures the likelihood that the genes are highly ranked in the universe of genes, as given by their association satatistics [149].

As previously mentioned, over representation analyses refer to the identification of gene-sets whose elements were mapped by significant *omic* variables on a higher rate than what would be expected by chance. These types of analyses are summarized in two steps. First, the list of significantly associated *omic* variables is identified and mapped to their corresponding genes. Second, we apply a statistical test, such as the hypergeometric test (one-tailed Fisher's exact test) to assess if the identified genes are likely to belong to a common predefined gene-set. The hypergeometric test follows an **urn model** in which a sequence of draws is performed, each of which with two possible outcomes (success or failure). m draws without replacement are performed from a possible total of N draws, in which there is n number of successes. The total number of drawn successes (k) follows a hypergeometric distribution. A hypergeometric test can be, for instance, applied on a transcriptome-wide analysis, in which we have mapped n genes, with differentially expressed (DE) transcription, from a total of N genes tested. The significant enrichment of a GO term with m genes, k of which were found DE, can be computed for the following table of parameters

	DE	non-DE	total
inside gene set	k	$m - k$	m
outside gene set	$n - k$	$N + k - n - m$	$N - m$
total	n	$N - n$	N

which define the hypergeometric distribution of the random variable X representing the probability of getting exactly k DE genes in the GO term

$$\Pr(X = k) = \frac{\binom{m}{k}\binom{N-m}{n-k}}{\binom{N}{n}}. \tag{10.1}$$

The probability of observing k or more genes inside a particular gene set is

$$\Pr(X >= k) = \sum_{x=k}^{n} \frac{\binom{m}{x}\binom{N-m}{n-x}}{\binom{N}{n}}, \tag{10.2}$$

which is taken as the enrichment measurement of the GO term. The lower the value the less likely that we observe k differentially expressed (DE) genes in the GO term by chance and thus the more the gene set is enriched in DE genes.

Consider an example in which 2,671 genes (N) have been tested for differential expression between two sample conditions and 89 genes were found DE (n). Among the DE genes, 18 (k) are annotated to a specific functional gene set (or pathway), which contains in total 170 genes (m). We want to assess whether the gene-set is over represented (i.e. enriched) with DE genes. For this purpose, we apply Equations 10.4 as follows:

```
> N <- 2671
> m <- 170
> n <- 89
> k <- 18
> sum(choose(m, k:m) * choose(N-m, n-(k:m)) / choose(N, n))
[1] 6.887546e-06
```

We observe that the probability of finding 18 or more genes out of 89 in the gene set containing 170 genes is very small. In particular it is lower than 0.05. Therefore, we can conclude that our list of DE genes are over represented in that gene set. This setup also corresponds to the following 2×2 contingency table. Therefore, the previous probability is equivalent to the P-value of a one-tailed Fisher's exact test:

```
> t <- array(c(k,n-k,m-k,N+k-n-m), dim=c(2,2),
+             dimnames=list(GS=c("in","out"),
+                           DE=c("yes","no")))
> t
     DE
GS    yes   no
  in   18  152
  out  71 2430
> fisher.test(t, alternative="greater")

Fisher's Exact Test for Count Data

data:  t
p-value = 6.888e-06
alternative hypothesis: true odds ratio is greater than 1
95 percent confidence interval:
 2.436118      Inf
sample estimates:
odds ratio
  4.049478
```

By using this test we can provide not only the p-value of over representation, but also the odds ratio of enrichment. In that case, one could conclude that the DE genes are four times more present in that gene set than expected (OR=4.05).

The Bioconductor packages *Category* and *GOstats* perform hypergeometric tests to assess over- and under-representation of DE genes in gene sets.

We demonstrate the enrichment analysis for the results obtained for the transcriptomic-wide analysis of the Alzheimer's disease study, discussed in Chapter 7. The complete transcriptome-wide analysis is reproduced by the following code

```
> library(GEOquery)
> library(sva)
> library(limma)


> #load transcriptomic data
> gsm.expr <- getGEO("GSE63061")
> gsm.expr <- gsm.expr[[1]]
> exprGEO <- exprs(gsm.expr)
>
> #get phenotype data
> pheno <- pData(phenoData(gsm.expr))
> status <- pheno$characteristics_ch1
> status <- gsub("status: ","", as.character(status))
> fstatus <- factor(status)
> levels(fstatus)<-gsub(" ", "", levels(fstatus))
>
> #select variables
> age <- substr(pheno$characteristics_ch1.2, 6,7)
> age<-as.numeric(age)
> sex <- pheno$characteristics_ch1.3
> phenodat<-data.frame(fstatus, age, sex)
>
> #build models
> mod0 <- model.matrix( ~ age+sex, data = phenodat)
> mod <- model.matrix( ~ fstatus+age+sex, data = phenodat)
>
> #compute surrogate variables for batch effects
> svobj <- sva(exprGEO, mod, mod0, n.sv=2)
> design <- model.matrix(~ 0+fstatus+sex+age+svobj$sv[,1]+svobj$sv[,2])
> colnames(design) <- c(levels(fstatus),"age","sex", "sva1","sva2")
>
> #fit the model for the desired contrast
> fit <- lmFit(exprGEO, design)
> contrast.matrix <- makeContrasts(AD-CTL, MCI-CTL, AD-MCI, levels=design)
>
> fit2 <- contrasts.fit(fit, contrast.matrix)
> fit2 <- eBayes(fit2)
```

The object `fit2` contains the results of the transcriptome-wise association analysis. For enrichment analysis, we first obtain the P-values of association for the contrast of Alzheimer's disease against controls. These are the associations for each of the Illumina probes

```
> results<-fit2$p.value
> head(results)
              Contrasts
                 AD - CTL   MCI - CTL    AD - MCI
  ILMN_1343291 0.0001343099 0.23838365 0.012397242
```

```
    ILMN_1343295 0.1349217169 0.46002556 0.492158315
    ILMN_1651209 0.1706294304 0.27616550 0.836459962
    ILMN_1651210 0.2716501412 0.02707706 0.225797561
    ILMN_1651221 0.4511555267 0.26407231 0.675401774
    ILMN_1651228 0.0171718691 0.64695073 0.005701243
>
> pvalues<-results[,"AD - CTL"]
> head(pvalues)
ILMN_1343291 ILMN_1343295 ILMN_1651209 ILMN_1651210 ILMN_1651221
0.0001343099 0.1349217169 0.1706294304 0.2716501412 0.4511555267
ILMN_1651228
0.0171718691
```

We then select the Illumina probes that are significantly associated to the case/control differences between individuals after considering multiple comparisons (FDR at 1% level)

```
> pAdj <- p.adjust(pvalues, "fdr")
> psig <- pAdj[pAdj < 0.01]
> head(psig)
ILMN_1343291 ILMN_1651254 ILMN_1651776 ILMN_1652073 ILMN_1652085
6.176532e-03 6.650025e-06 7.517934e-04 1.013183e-10 1.479804e-03
ILMN_1652806
7.091283e-05
```

This filter provides a list of 819 DE probes at 1%-FDR level. The mapping for the probes to the genes is provided in the annotation of the ExpressionSet and retrieved with (fData)

```
> genesIDs <- as.character(fData(gsm.expr)$ILMN_Gene)
> names(genesIDs)<-rownames(gsm.expr)
>
> psigAnnot<-data.frame(genesIDs=genesIDs[names(psig)],
+                        pvalAdj=psig,stringsAsFactors=FALSE)
>
> head(psigAnnot)
                genesIDs      pvalAdj
ILMN_1343291     EEF1A1 6.176532e-03
ILMN_1651254        LPP 6.650025e-06
ILMN_1651776      FHOD1 7.517934e-04
ILMN_1652073 LOC653658 1.013183e-10
ILMN_1652085 MPHOSPH10 1.479804e-03
ILMN_1652806      ATP5J 7.091283e-05
```

The gene names provided in the ExpressionSet object are the gene symbols. However, enrichment analyses are performed with *GOsats* with the ENTREZID identifier. Using the org.Hs.eg.db database, discussed before, we retrieve all the gene symbols and the entrez identifiers annotated in the human genome.

```
> genes <- keys(org.Hs.eg.db, keytype="SYMBOL")
>
> geneUniverse <- select(org.Hs.eg.db, keys = genes,
```

```
+                          columns=c("ENTREZID"), keytype = "SYMBOL")
>
> head(geneUniverse)
  SYMBOL ENTREZID
1   A1BG        1
2    A2M        2
3  A2MP1        3
4   NAT1        9
5   NAT2       10
6   NATP       11
```

We obtain 60140 genes, which is the number of rows in the `geneUniverse`. The entrez identifiers for the 819 genes with DE transcripts is obtained from

```
> mappedgenes <- psigAnnot$genesIDs
> mappedgenes <- intersect(mappedgenes,geneUniverse$SYMBOL)
> selmappedgenes <- geneUniverse$SYMBOL%in%mappedgenes
> mappedgenesIds <- geneUniverse$ENTREZID[selmappedgenes]
> head(mappedgenesIds)
[1] "20"  "34"  "37"  "101" "199" "204"
```

This is a list of 514 genes. The enrichment analysis is set up by a `GOHyperGParams` to be passed to the `try(hyperGTest)`. The list of parameters include the list of genes with significant DE (`mappedgenesIds`), the total set of genes (`geneUniverse$ENTREZID`), the annotation data base (`org.Hs.eg.db`), the type of ontology (Biological process-BP), the *P*-value cutoff and the direction of the enrichment (over-representation of significant DE genes)

```
> library(GOstats)
>
> params <- new("GOHyperGParams", geneIds=mappedgenesIds,
+               universeGeneIds=geneUniverse$ENTREZID,
+               annotation="org.Hs.eg.db", ontology="BP",
+               pvalueCutoff=0.05, conditional=FALSE,
+               testDirection="over")
```

Enrichment tests are performed for all annotated GO terms using the `hyperGTest` function

```
> hgOver <- hyperGTest(params)
> hgOver
Gene to GO BP  test for over-representation
4618 GO BP ids tested (679 have p < 0.05)
Selected gene set size: 479
   Gene universe size: 17653
   Annotation package: org.Hs.eg
> summary(hgOver)[c(1:5),]
      GOBPID       Pvalue OddsRatio  ExpCount Count Size
1 GO:0006614 1.657370e-23  17.78667  2.469212    29   91
2 GO:0006613 9.255554e-23  16.45449  2.604883    29   96
3 GO:0016071 1.972136e-22   4.27150 22.412848    78  826
4 GO:0045047 4.623231e-22  15.30735  2.740554    29  101
5 GO:0072599 1.558404e-21  14.49830  2.849091    29  105
                                                     Term
```

```
1     SRP-dependent cotranslational protein targeting to membrane
2                   cotranslational protein targeting to membrane
3                                         mRNA metabolic process
4                                         protein targeting to ER
5 establishment of protein localization to endoplasmic reticulum
```

Note that gene sets forming GO terms are not disjoint and overlapping GO terms will have correlating enrichment. We then consider using a conditional test and generate another report:

```
> conditional(params) <- TRUE
> hgOverCond <- hyperGTest(params)
> summary(hgOver)[c(1:3),]
     GOBPID        Pvalue OddsRatio  ExpCount Count Size
1 GO:0006614 1.657370e-23  17.78667  2.469212    29   91
2 GO:0006613 9.255554e-23  16.45449  2.604883    29   96
3 GO:0016071 1.972136e-22   4.27150 22.412848    78  826
                                                     Term
1 SRP-dependent cotranslational protein targeting to membrane
2                 cotranslational protein targeting to membrane
3                                       mRNA metabolic process
```

A full report can be rendered in HTML format

```
> htmlReport(hgOverCond, file="goBPcond.html")
```

We thus see a strong enrichment in GO terms relating the function of the endoplasmatic reticulum with a well-documented role in Alzheimer's disease [81].

Further insight can be gained from using the KEGGdatabase (http://www.genome.jp/kegg/). In the KEGG database, numerous biochemical pathways and reactions have been curated from the literature. The pathways represent biochemical cellular interactions with identified functions within metabolic, cellular, genetic and environmental information processes, organismal systems, human diseases and drug development. The database is accessed in Bioconductor with the *KEGG.db* library. In the *org.Hs.eg.db* database, the dataset org.Hs.egPATH contains the mapping of human genes (with entrez identifiers) to the KEGG pathways identifiers, which can be directly accessed with (toTable) and formatted into a *data.frame*

```
> #?org.Hs.egPATH
> frame <- toTable(org.Hs.egPATH)
> keggframeData <- data.frame(frame$path_id, frame$gene_id)
> head(keggframeData)
  frame.path_id frame.gene_id
1        04610             2
2        00232             9
3        00983             9
4        01100             9
5        00232            10
6        00983            10
```

The package *GSEABase* allows flexible implementation of gene-sets collections that can be passed to the enrichment analysis. When using non-traditional sources about KEGG, the gene-set collections are safely annotated against the *KEGG.db* library

```
> library(GSEABase)
> library(KEGG.db)
>
> keggFrame <- KEGGFrame(keggframeData, organism="Homo sapiens")
> gsc.KEGG <- GeneSetCollection(keggFrame, setType = KEGGCollection())
>
> gsc.KEGG
GeneSetCollection
  names: 00010, 00020, ..., 05416 (229 total)
  unique identifiers: 124, 125, ..., 1525 (5869 total)
  types in collection:
    geneIdType: KEGGFrameIdentifier (1 total)
    collectionType: KEGGCollection (1 total)
```

We thus create a gene-set collection of 229 KEGG pathways with 5869 genes. We can then perform an enrichment analysis for each of these pathways, defining a new set of parameters that will be passed to the hypergeometric test function **hyperGTest**

```
> KEGG.params.bp <-  GSEAKEGGHyperGParams(name="KEGG",
+                  geneSetCollection=gsc.KEGG,
+                  geneIds=mappedgenesIds,
+                  universeGeneIds=geneUniverse$ENTREZID,
+                  pvalueCutoff=0.05,
+                  testDirection="over")
>
> KEGG.results.bp <- hyperGTest(KEGG.params.bp)
> head(summary(KEGG.results.bp))
  KEGGID       Pvalue OddsRatio ExpCount Count Size
1 03010 7.033744e-21 14.006357 3.504174    30   91
2 00190 2.691936e-12  6.790660 5.082978    26  132
3 03040 2.900516e-07  4.704106 4.890441    19  127
4 05010 1.329892e-06  3.856899 6.430738    21  167
5 05016 1.594308e-06  3.672025 7.046856    22  183
6 05012 1.936332e-06  4.273609 5.005964    18  130
                      Term
1                  Ribosome
2 Oxidative phosphorylation
3               Spliceosome
4       Alzheimer's disease
5      Huntington's disease
6      Parkinson's disease
```

We see that few KEGG pathways are significantly enriched in transcriptomic differences between Alzheimer's cases and controls. In particular, we have identified 11 genes from 167 that take part in the disease, making the Alzheimer's disease pathway (05010) significantly enriched. Note in addition that other interesting pathways have been identified, such as those related to oxidative phosphorylation and other neurodegenerative diseases.

The `hyperGTest` function allows performing the hypergeometric test in gene-sets that can be defined by the user, and *GSEABase* allows building general gene collections and verifies KEGG pathways against *KEGG.db*. However, there are other Bioconductor packages that are quick to use, such as *cluster-Profiler*. This package internally maps the genes with its version of KEGG and GO pathways allowing fast testing of enrichment analysis. The package is also strong on the graphic representations of the results.

With `enrichKEGG` from *clusterProfiler* [177], we can obtain similar results as before, with less amount of code but less control

```
> library(clusterProfiler)
>
> res.enrich <- enrichKEGG(gene = mappedgenesIds,
+                organism = 'hsa',
+                pvalueCutoff = 0.05)
>
> res.enrich[1:5, 1:7]
                 ID                                  Description GeneRatio
hsa03010 hsa03010                                     Ribosome    37/265
hsa00190 hsa00190                      Oxidative phosphorylation    26/265
hsa04932 hsa04932 Non-alcoholic fatty liver disease (NAFLD)     21/265
hsa03040 hsa03040                                  Spliceosome    19/265
hsa05010 hsa05010                            Alzheimer disease    21/265
            BgRatio        pvalue      p.adjust        qvalue
hsa03010 153/7852 4.072384e-22 9.895893e-20 9.473651e-20
hsa00190 133/7852 1.938655e-13 2.355466e-11 2.254962e-11
hsa04932 149/7852 2.260664e-08 1.831138e-06 1.753006e-06
hsa03040 134/7852 9.816998e-08 5.963827e-06 5.709360e-06
hsa05010 171/7852 2.548067e-07 1.238361e-05 1.185522e-05
```

Several plots are available to explore the results. They include a dotplot (Figure 10.1) that shows the number of genes mapped in each of the pathways

```
> dotplot(res.enrich)
```

The package also contains other functions such as `enrichDAVID` to perform enrichment analysis using DAVID [61] or `enrichDO` to perform a Disease Ontology analysis. Similar to *GSEABase* package, *clusterProfiler* contains a general function called `enricher` which facilitates any enrichment analysis considering other gene sets as those available at MSigDB database (`http://software.broadinstitute.org/gsea/msigdb`). Let us assume we want to perform an enrichment analysis of hallmark gene sets which are availalbe at the collection named 'H from MSigDB. Once the 'gmt file is downloaded, the enrichment analysis can be performed by:

```
> gset <- read.gmt("h.all.v6.2.entrez.gmt")
> res.enrich.H <- enricher(gene = mappedgenesIds,
+                          TERM2GENE = gset)
> res.enrich.H[1:5, 3:5]
                                   GeneRatio BgRatio        pvalue
HALLMARK_OXIDATIVE_PHOSPHORYLATION   29/177 200/4386 7.858178e-10
```

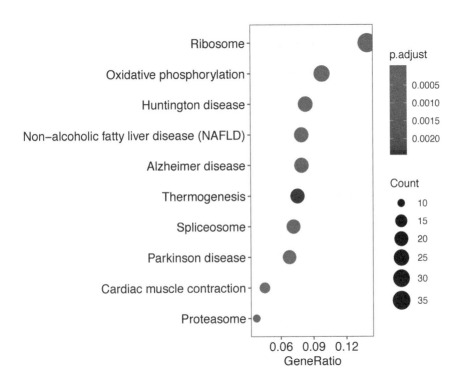

FIGURE 10.1
Dotplot for enrichment analysis of top DE genes obtained from transcriptomic data analysis of Alzheimer's disease.

```
HALLMARK_MYC_TARGETS_V1               22/177 200/4386 1.280537e-05
HALLMARK_DNA_REPAIR                   16/177 150/4386 3.000517e-04
HALLMARK_ALLOGRAFT_REJECTION          16/177 200/4386 6.284938e-03
HALLMARK_MTORC1_SIGNALING             14/177 200/4386 2.992623e-02
```

10.5   Overlap with functional genomic regions

In Chapter 5 we illustrate how to assess the association between genomic variants (CNVs, genetic mosaicims and inversions) and disease. In those situations, it is of interest to determine if the genomic variant contains functional genomic regions such as genes, promoters or enhancers. The *regioneR* package implements a general framework for testing the overlap of genomic regions with functional elements based on permutation sampling [39].

Let us assume, we want to assess the overlap between the significant CNV regions obtained in the example of BRCA data from TCGA described in Section 5.2.1 and protein-coding regions. We now recall the results we obtained in the association analysis

```
> brca.gr.sig
GRanges object with 7362 ranges and 3 metadata columns:
          seqnames              ranges strand |   counts
             <Rle>           <IRanges>  <Rle> | <matrix>
   [1]        chr1 149898951-150333087      * |     1:30
   [2]        chr1 150335347-151170789      * |     0:30
   [3]        chr1 154984468-155005476      * |     0:21
   [4]        chr1 155006954-155050185      * |     0:22
   [5]       chr11   60943071-61361689      * |     0:16
   ...         ...                 ...    ... .      ...
[7358]       chr11   69559601-69650887      * |     0:66
[7359]       chr11   70695866-71002410      * |     1:53
[7360]       chr11   73552530-73878688      * |     0:22
[7361]       chr11   74853813-74901601      * |     0:21
[7362]       chr16     653459-2025678      * |     1:16
                       pvalue                     BH
                    <numeric>              <numeric>
   [1] 9.17196804835805e-07  6.85696840142296e-06
   [2] 4.02900812259604e-08  3.69842635940347e-07
   [3] 6.60494300658908e-06  4.05859140379805e-05
   [4] 3.64039791063025e-06  2.34223634083979e-05
   [5] 0.000138664831551331  0.000654493148010495
   ...                  ...                    ...
[7358] 2.94813576696704e-17  9.70149048297054e-16
[7359] 1.24305369365935e-12  2.43117126979776e-11
[7360] 3.64039791063025e-06  2.34223634083979e-05
[7361] 6.60494300658908e-06  4.05859140379805e-05
[7362] 0.00101824759029954   0.00399167866271459
  -------
  seqinfo: 23 sequences from an unspecified genome; no seqlengths
```

To obtain the location of protein-coding genes, we query available Human annotation from Ensembl

```
> library(AnnotationHub)
> ah <- AnnotationHub()
> ahDb <- query(ah, pattern = c("Homo sapiens", "EnsDb"))
> ahDb
```

and retrieve gene coordinates in the UMD3.1 assembly (Ensembl 92).

```
> ahEdb <- ahDb[["AH60977"]]
> hg.genes <- genes(ahEdb)
```

We then need to verify that the `seqlevels` annotation of both `GenomicRanges` are the same:

```
> seqnames(hg.genes)
factor-Rle of length 65256 with 423 runs
  Lengths:                       5349 ...                       518
  Values :                       chr1 ...                      chrY
Levels(423): chr1 chr10 chr11 chr12 chr13 ... chrLRG_93 chrMT chrX chrY
> seqnames(brca.gr.sig)
factor-Rle of length 7362 with 1548 runs
  Lengths:      4      9      3     11      1 ...     21      1      2      8      1
  Values :   chr1  chr11  chr12  chr20  chr11 ...  chr11  chr17  chr10  chr11  chr16
Levels(23): chr1 chr2 chr3 chr10 chr14 ... chr16 chr19 chr13 chr22 chr18
```

otherwise we need to re-assign the `seqlevels`

```
> seqlevels(hg.genes) <- paste0("chr", seqlevels(hg.genes))
```

Let us run the enrichment in chromosome 1:

```
> sel.protein <- hg.genes[seqnames(hg.genes)%in%c("chr1")]
> sel.protein <- sel.protein[sel.protein$gene_biotype == "protein_coding"]
> sel.cnvs <- brca.gr.sig[seqnames(brca.gr.sig)%in%c("chr1")]
> length(sel.cnvs)
[1] 234
```

We then apply an overlap permutation test with 100 permutations (`ntimes=100`), while maintaining chromosomal distribution of the CNV region set (`per.chromosome=TRUE`). Setting the option `count.once=TRUE` counts an overlapping CNV region only once. To randomly sample regions from the entire genome no masking is required (`mask=NA`) and to draw robust conclusions a minimum of 1000 permutations should be performed

```
> library(regioneR)
> library(BSgenome.Hsapiens.UCSC.hg19.masked)
> res.prot <- overlapPermTest(A=sel.cnvs, B=sel.protein,
+                             mask=NA, genome="hg19", ntimes=100,
+                             per.chromosome=TRUE, count.once=TRUE)
> res.prot
$numOverlaps
P-value: 0.0099009900990099
Z-score: 7.6781
Number of iterations: 100
Alternative: greater
Evaluation of the original region set: 220
Evaluation function: numOverlaps
Randomization function: randomizeRegions

attr(,"class")
[1] "permTestResultsList"
```

The permutation P-value indicates a statistically significant overlaps of CNVs with protein coding regions. From 234 CNVs, 220 overlap with at least one protein coding gene. We can also repeat the analysis for promoters from UCSC defined as -2000 to +200 base pairs to the TSS according to the genome assembly hg19. Analyzing chromosome 1

```
> file.prom <- "http://gattaca.imppc.org/regioner/data/UCSC.promoters.hg19.bed"
> promoters <- toGRanges(file.prom)
> sel.prom <- promoters[seqnames(promoters) %in% c("chr1")]
```

we obtain

```
> res.prom <- overlapPermTest(sel.cnvs, sel.prom,
+                             ntimes=100, genome="hg19",
+                             count.once=TRUE)
> res.prom
$numOverlaps
P-value: 0.0099009900990099
Z-score: 42.6637
Number of iterations: 100
Alternative: greater
Evaluation of the original region set: 184
Evaluation function: numOverlaps
Randomization function: randomizeRegions

attr(,"class")
[1] "permTestResultsList"
```

We also observe significant overlap of CNVs in regions with promoters. Notice that the function **permTest** allows user-defined functions for randomizing control and for evaluating additional genomic measures.

10.6   Chemical and environmental enrichment

CTDquerier is a Bioconductors package that provides novel hypotheses about the relationships between chemicals, genes and diseases [53]. It uses data from Comparative Toxicogenomics Database (*CTD*) which provides information about the disease-gene-chemical triad [53]. For instance, we can use *CTD-querier* to provide curated evidence for our list of genes relating to Alzheimer's disease, chemical interactions and other diseases.

A total of 514 genes are considered to be DE between Alzheimer's disease and control patients (object **mappedgenes**):

```
> length(mappedgenes)
[1] 514
```

We start inquiring how many of those genes are annotated in CTD. The name of the genes in Gene Symbol format is required **mappedgenes**.

```
> library( CTDquerier )
```

```
> genesAD <- query_ctd_gene( terms = mappedgenes,
+                            verbose = TRUE )
```

Figure 10.2 describes the number of genes that are found in CTD

```
> library( ggplot2 )
> plot( genesAD ) + ggtitle( "Number of genes" )
```

The genes that are not found in CTD are listed below

```
> get_terms(genesAD)[[ "lost" ]]
[1] "LOC285074"    "LOC100131859" "LOC100132287" "LOC728877"
[5] "LOC100129118" "LOC100132658" "LOC339192"
```

The call to **query_ctd_gene** function downloads "all the available information" in *CTD* related to our gene list. This object provides the following information:

```
> genesAD
Object of class 'CTDdata'
-------------------------
 . Type: GENE
 . Length: 507
 . Items: EEF1A1, ..., PABPC1L
 . Diseases: 3047 ( 278100 / 583640 )
 . Gene-gene interactions: 34161 ( 45431 )
 . Gene-chemical interactions: 28594 ( 41672 )
 . KEGG pathways: 7119 ( 7119 )
 . GO terms: 10696 ( 10706 )
```

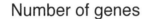

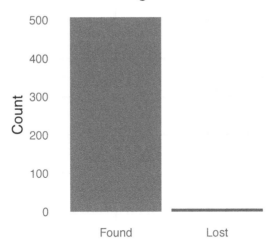

FIGURE 10.2
Genes found in CTD that were obtained from the analysis of GEO data GSE63061. Seven from 514 are not present in CTD.

The line starting with **#Diseases** indicates the number of relationships between a gene (or chemical when using **query_ctd_chem** function) and diseases that are annotated in *CTD*. For our list of genes, there are a total of 583640 relations (both inferred and curated) between our genes of interest and diseases. The table with the relations between genes and diseases can be obtained using the method **get_table**.

```
> diseases <- get_table(genesAD, index_name = "diseases" )
> colnames(diseases)
[1] "Disease.Name"      "Disease.ID"       "Direct.Evidence"
[4] "Inference.Network" "Inference.Score"   "Reference.Count"
[7] "GeneSymbol"        "GeneID"
> dim(diseases)
[1] 583640       8
```

We can list the diseases that have been linked to our gene list

```
> length( unique(diseases$Disease.Name))
[1] 3047
```

Note that diseases can be annotated in a very specific way, and hence, a large number of unique diseases can be obtained. The column **Direct.Evidence** provides information about the type of relationship observed in the literature (see **http://ctdbase.org/**). In our case:

```
> table(diseases$Direct.Evidence)

                                                marker/mechanism
                              277449                         617
    marker/mechanism|therapeutic              therapeutic
                                   4                          30
```

This indicates that there are 651 pieces of direct evidence in the literature showing that within our gene list there are relevant associations with disease biomarker or etiology of diseases. Disease relationships can be obtained by

```
> mask <- !is.na(diseases$Direct.Evidence) &
+           diseases$Direct.Evidence != ""
> diseases_cu <- diseases[mask,]
> dim(diseases_cu )
[1] 651    8
> length(unique(diseases_cu$Disease.Name))
[1] 368
```

There are 368 unique diseases linked to our list of genes, supported by 651, 8 curated relationships. CTD highlights associations obtained from direct evidence. We show how to retrieve this information.

We subset `"Alzheimer Disease"` from the table `diseases`, which provides all the relationships between genes and diseases. We extract the top genes mostly associated with Alzheimer's disease in *CTD*

```
> adgenes <- diseases[diseases$Disease.Name == "Alzheimer Disease" ,]
> o <- order(as.numeric(adgenes$Inference.Score), decreasing = TRUE)
> adgenes.o <- adgenes[o, c(1,5:8)]
> head(adgenes.o)
DataFrame with 6 rows and 5 columns
        Disease.Name Inference.Score Reference.Count  GeneSymbol
         <character>     <character>       <integer> <character>
1 Alzheimer Disease           94.24              85         BAX
2 Alzheimer Disease           83.54              79        SOD1
3 Alzheimer Disease           61.56              44       NR3C1
4 Alzheimer Disease           54.04              47       CREBBP
5 Alzheimer Disease           53.36              58       ITGAM
6 Alzheimer Disease           51.15              39       CASP1
        GeneID
   <character>
1          581
2         6647
3         2908
4         1387
5         3684
6          834
```

We can also extract the list of genes associated with air pollutants chemicals *CTD*. We obtain the information about air pollutants from *CTD*, using the function `query_ctd_chem`

```
> air <- query_ctd_chem( terms = "Air Pollutants" )
> air
Object of class 'CTDdata'
------------------------
 . Type: CHEMICAL
 . Length: 1
 . Items: AIR POLLUTANTS
 . Diseases: 1507 ( 105 / 2175 )
 . Diseases: 1507 ( 2175 )
 . Chemical-gene interactions: 1692 ( 1873 )
 . KEGG pathways: 374 ( 374 )
 . GO terms: 1875 ( 1875 )
```

We then perform an enrichment analysis with the function **enrich**, that requires the gene universe. We use the **geneUniverse** but a list is also available within *CTDquerier*

```
> hgnc_universe <- read.delim(system.file( "extdata", "HGNC_Genes.tsv",
+                                           package="CTDquerier" ),
+                             sep = "\t", stringsAsFactor = FALSE )
```

HGNC_Genes.tsv corresponds to genes associated with air pollutants.

```
> ans.air <- enrich( genesAD, air,
+                    universe = hgnc_universe$Approved.Symbol,
+                    use = "all" )
> ans.air

Fisher's Exact Test for Count Data

data:   table(universe %in% y, universe %in% x)
p-value = 4.663e-10
alternative hypothesis: true odds ratio is not equal to 1
95 percent confidence interval:
 2.097080 3.860931
sample estimates:
odds ratio
  2.871975
```

We then observe that our list of genes, DE in Alzheimer's disease, are enriched in genes associated with air pollutants (OR = 2.9 , *P*-value = 4.663e-10).

11

Multiomic data analysis

CONTENTS

11.1   Chapter overview

The objective of the present chapter is to introduce some approaches to analyze studies with multiomic data. These are studies that, for instance, have collected genomic, transcriptomic and exposomic data, or other types of *omic* data, on the same individuals. The volume of the data allows for multiple and complex relationships within and between the datasets. The broad objective is therefore to identify the relationships that are derived from underlying biological mechanisms. The field is wide open and while many types of analysis can be applied, and will be developed, we keep an eye on those that allow us to explain phenotypic differences between individuals. We demonstrate specific examples comprising two different approaches, namely multi-staged and meta-dimensional analyses.

11.2   Multiomic data

Understanding the molecular basis of complex traits is an open question. Recent advances in technology have allowed exploring the complexity of the problem by incorporating multiple pieces of information such as molecular *omic* data, body imaging, clinical data, and measurements of environmental exposures, among others. Therefore, one of the current challenges is to perform valid biological inferences from the vast amount of data that is being generated. In statistical terms the challenge may be tackled from the integration of multiple sources of information, encoded in different tables. In this chapter, we will demonstrate some integration methods in the context of *omic* data.

Figure 11.1 illustrates different types of data that may be collected in a hypothetical study. Before looking at the inter-relations among datasets, it is worth noticing that within datasets there is already large complexity, in particular, if they are used to infer different biological entities from the ones intended. For instance, SNP data can be used to detect mosaicisms or inversion polymorphisms, RNA-seq data may be used for to infer either gene expression or alternative splicing abundance, or functional neuroimaging can be used to study brain activation in specific regions or to infer brain connectivity.

Data integration, also known as integrative bioinformatics, integrated analysis, crossomics, multi-dataset analysis or data fusion, includes the computational combination of datasets or the joint analysis of different tables, from different measurement modalities. The main aim of those analyses is to describe patterns in the data that correspond to underlying biological processes.

11.3   Massive pair-wise analyses between *omic* datasets

The first approach to integrating two *omic* datasets is perhaps to correlate all variables in one *omic* dataset with all variables in the second dataset. For high-dimensional data, the approach is clearly inefficient in terms of statistical power, as the threshold of significance should be dropped by an order of magnitude that is the product between the dimensions of the *omic* datasets. If the number of features for the tables are dim1 and dim2 then the significance level is dropped to $0.05/10^{dim1*dim2}$. In this context, this type of analysis benefits from predefining features or regions of interest.

Massive pair-wise analyses are supported by the *MEAL* package. For instance, we can perform association tests between methylation and gene expression in a given region of interest, using the methods described in [143] and implemented in the `correlationMethExprs` function. For an analysis of this type, methylomic and transcriptomic variables are paired by proximity. Tran-

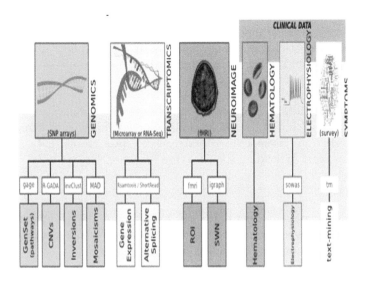

FIGURE 11.1
Hypothetical example illustrates the multivariate nature of biomedical studies. It shows different types of *omic* data that can be collected in current studies (genomic, transcriptomic, neuroimage, clinical) and how they also provide more information than the direct measurement of primary features. Some R/Bioconductor packages are also shown for the analysis of the data R.

scription variables are paired to a CpG if they are completely inside a range of $\pm$ 250Kb from the CpG. The distance in base-pair units can be changed with the parameter `flank`. The correlation between methylation and transcription is done by a linear regression. To account for technical (e.g. batch) or biological (e.g. sex, age) artifacts, a model for K covariates (z) is fitted for each *omic* variable:

$$x_{ij} = \sum_{k=1}^{K} \beta_{ik} z_{kj} + r_{ij}, i = 1, ..., P$$

where x_{ij} is the methylation or transcription variable i for individual j and β_{ik} is the effect of covariate z_{kj} ($k = 1...K$). r_{ij} are the residuals which are used to assess the correlation between different types of *omic* data.

We illustrate this type of analysis using the breast cancer study with multi-*omic* data from TCGA [101] introduced in Section 3.9. Data is available through the *brgedata* package. The object is of *MultiDataSet* class containing miRNA, miRNAprecursor, RNAseq, methylation, proteins from a RPPA array, and GISTIC SNP calls (CNA and LOH). Clinical data are also available using `pData` of any of the components. Our main aim is to determine those features that are associated with estrogen receptor status (ER+ vs ER-).

Let us start by loading the data into R

```
> data(breastMulti, package="brgedata")
```

The association analysis between transcriptomic and methylomic data only considers probes in a target region. Data for the region is obtained passing a `GenomicRanges` on the third index of an array selector [:

```
> library(Biobase)
> library(GenomicRanges)
> targetRange <- GRanges("chr19:350000-45550000")
> breastMulti.target <- breastMulti[, , targetRange]
```

The number of features of each *omic* table that mapped over the selected genomic range is

```
> dims(breastMulti.target)
$methylation
Features  Samples
      20       79

$expression
Features  Samples
     316       79
```

The correlation between the methylomic and transcriptomic variables in the selected region is computed with the `correlationMethExprs` function from *MEAL*

```
> library(MEAL)
> methExprs <- correlationMethExprs(breastMulti.target)
> head(methExprs)
        cpg  exprs      Beta       se     P.Value    adj.P.Val
1 cg07753644  ICAM1 -2.969179 0.5953793 3.696990e-06 0.0003733960
2 cg19996355   PBX4 -2.369049 0.5050955 1.160150e-05 0.0003967379
3 cg18133957  REEP6  3.964528 0.8460039 1.178429e-05 0.0003967379
4 cg04103514 SLC1A6  4.375455 0.9712205 2.328168e-05 0.0005878625
5 cg07753644  PDE4A  1.843641 0.4293005 5.051752e-05 0.0010204540
6 cg16992787   CAPS -3.581154 0.9385986 2.728147e-04 0.0045923805
```

The results show the CpG names, the expression probes (gene symbol), the effects of the association and with the standard errors, the *P*-values of association and the adjusted *P*-values, using the Benjamini-Hochberg correction. Results, ordered by the adjusted *P*-values, show several significant associations between different CpGs and transcript levels.

A cis- transcriptome analysis can also be carried out for significant CpGs associated with a phenotype or grouping variable. In our case, when we compare ER+ vs ER- we do not obtain any DMPs that pass multiple comparisons correction, but the analysis can be performed, for instance, with the top-10 CpGs as follows

```
> methy <- breastMulti[["methylation"]]
> resER <- runPipeline(methy, variable_names = "ER.Status",
+                      sva=TRUE)
> dmps <- getProbeResults(resER)
> head(dmps)
                logFC       CI.L        CI.R     AveExpr          t
cg24332422  0.4284006  0.20809316  0.6487081 0.22341696   3.872200
cg01462829  0.1893517  0.06789438  0.3108090 0.07135152   3.104434
cg17525406  0.4552882  0.16243246  0.7481439 0.35030441   3.095776
cg02250594  0.4044269  0.14332688  0.6655269 0.47392884   3.084396
cg18059088  0.2478118  0.08054732  0.4150764 0.11683830   2.950225
cg16501028 -0.2689283 -0.45158392 -0.0862727 0.68282623  -2.931841
              P.Value  adj.P.Val           B          SE chromosome
cg24332422 0.0002249271 0.1291081 -0.2953954 0.11566769       chr5
cg01462829 0.0026683859 0.4065825 -1.9392362 0.13104084       chr4
cg17525406 0.0027385219 0.4065825 -1.9563732 0.14706753       chr1
cg02250594 0.0028333275 0.4065825 -1.9788476 0.09371079      chr18
cg18059088 0.0042063960 0.4244262 -2.2392737 0.12274455       chr4
cg16501028 0.0044365110 0.4244262 -2.2742956 0.13148360      chr11
                 start
cg24332422 112073686
cg01462829  11430903
cg17525406   4715520
cg02250594  55103353
cg18059088  11430353
cg16501028  32450000
>
> cpgs <- rownames(dmps)[1:10] #select top-10 CpGs
> methExprs <- correlationMethExprs(breastMulti,
+                                   sel_cpgs=cpgs)
> methExprs
```

```
           cpg     exprs       Beta        se      P.Value    adj.P.Val
1   cg19996355      PBX4 -2.3690488 0.5050955 0.0000116015 0.0001276165
2   cg16501028       WT1  3.1750865 1.2515287 0.0132039462 0.0726217038
3   cg17108819      CD8B -0.8588676 0.4198625 0.0442118483 0.1621101103
4   cg02250594    CCDC68  0.7405556 0.4262180 0.0862963894 0.2373150707
5   cg24989962      STYX -0.4239871 0.2777992 0.1310484796 0.2883066551
6   cg19996355     CILP2  0.4350775 0.3312770 0.1929705301 0.3537793052
7   cg24989962   GPR137C -0.5260983 0.4928951 0.2891427368 0.3975712631
8   cg19996355     ZNF14  0.3317182 0.2909800 0.2578185194 0.3975712631
9   cg19996355    ZNF101 -0.2312611 0.3121282 0.4609973407 0.5634411942
10  cg18059088     HS3ST1 -0.3771166 0.8828755 0.6704651360 0.7375116496
11  cg01462829     HS3ST1  0.1903034 1.3556252 0.8887258901 0.8887258901
```

Here we observe that CpG cg19996355 is negatively associated with *PBX4* transcription (adjusted *P*-value 0.0001276). The results can be visuallized in Figure 11.2.

```
> library(vioplot)
> par(mfrow=c(1,2))
> cpg <- assay(methy)["cg19996355",]
> boxplot(cpg[methy$ER.Status=="Positive"],
+         cpg[methy$ER.Status=="Negative"],
+         col="gray80", names=c("Positive", "Negative"),
+         ylab="cg19996355", las=2)
>
> geneExpr <- breastMulti[["expression"]]
> gene <- assay(geneExpr)["PBX4",]
> plot(cpg, gene, xlab="cg19996355",
+      ylab="PBX4", type="n")
> points(cpg, gene, pch=21, bg="gray90")
> abline(lm(gene ~ cpg), lwd=2)
```

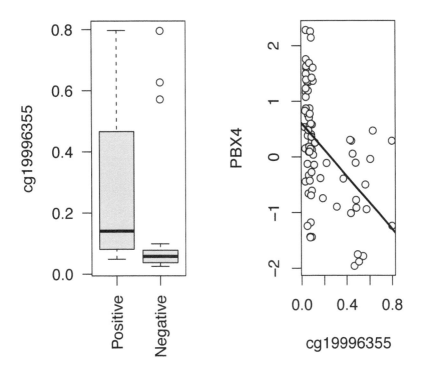

FIGURE 11.2
Significant correlation between the top CpG, associated with ER status, and
gene transcription from breast TCGA data.

11.4   Multiple-*omic* data integration

One main objectives is to combine *omic* datasets to obtain predictor variables that can explain phenotypic differences among individuals [123], with the hope that the combination is able to increase the level of explanation that may be obtained from a single data modality. In the search of predictor variables, two main approaches to data integration can be followed

- **Multi-staged analysis**: Involves integrating information using a stepwise or hierarchical analysis approach, such as genomic variation and domain-knowledge.

- **Meta-dimensional analysis**: Refers to the integration of different data types to build a multivariate model associated with a given outcome, such as multivariate analysis.

The next sections describe how to perform genomic variation, domain-knowledge and multivariate analyses of multiomic studies.

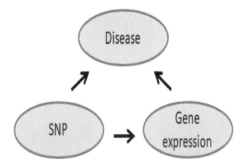

FIGURE 11.3
Hypothetical example illustrating the triangle approach in genomics where genomic, transcriptomic and disease data are integrated.

11.4.1 Multi-staged analysis

The multi-stage approach is based on two different strategies: genomic variation and domain-knowledge analyses. Figure 11.3 illustrates a hypothetical example of genomic variation analysis, also known as the triangle approach. It corresponds to the analysis of genomic, transcriptomic and phenotype data, in which SNPs with statistically significant association with the trait are firstly identified (Chapter 4); then the selected SNPs are tested for significant transcriptomic effects, e.g. eQTL - expression quantitative trait loci - effects and; finally transcripts with significant eQTL effects are tested for association with the trait.

11.4.1.1 Genomic variation analysis

When we have established significant associations between two types of *omic* variables with a phenotype, we may be interested in assessing the nature of their relationships, that is, the extent to which one variable may modify the association between the other variable and the phenotype. Identification of mediator variables can clarify the nature of the relationship between two variables [165]. Causal inference test (CIT) [95] and mediation analyses [165] are statistical techniques to identify potential mediators.

We illustrate CIT and mediation analysis for the effect of loss of chromosome Y (LOY) on cancer risk. Studies have shown individual's LOY is associated with cancer risk. We hypothesize that the independent variable LOY (X) is linked to the dependent variable cancer (Y) by:

$$X \rightarrow Y$$

One can assume, however, that LOY is not the real reason why cancer risk increases. We thus hypothesize that the relationship is mediated by the gene expression of a gene (M), that has been down-regulated by LOY. Therefore, we want to test whether LOY down-regulates gene expression which in turn increases cancer risk:

$$X \to M \to Y$$

This type of hypothesis is supported from the literature. For instance, down-regulation $DDX3Y$ in chromosome Y reduces cell proliferation which could lead to tumor growth and cancer.

CIT analysis is used to test the causal relationships among three variables. In our example, we use CIT analysis to test whether LOY status and transcription changes of $DDX3Y$ independently associate with cancer or if LOY changes transcript levels, which in turn modify cancer risk. CIT is based on hypothesis testing rather than effect estimation [95]. To illustrate CIT analysis, we use TCGA data for lung squamous cell carcinoma for which we have obtained LOY status using the *MADloy* package, as described in Chapter 5. The gene expression and tumor status (normal/cancer) have been obtained from the *RTCGA* package. All this information has been encapsulated in a *data.frame* called `lusc` that is available from the *brgedata* package.

```
> data(lusc, package="brgedata")
```

The R package *cit* implements the CIT method. There are two main functions `cit.bp` and `cit.cp` that are used to perform CIT for binary and continuous outcomes, respectively. In our case, we aim to test whether gene expression of $DDX3Y$ mediates the association between LOY and cancer. The analysis is simply performed by executing

```
> library(cit)
> L <- lusc[,"LOY"]              # instrumental variable
> G <- lusc[, "DDX3Y"]          # mediator
> T <- lusc[, "Cancer"]         # outcome
> C <- lusc[, "age"]            # confounders
> cit.bp(L, G, T, C, rseed=1234)
          p_cit    p_TassocL p_TassocGgvnL p_GassocLgvnT p_LindTgvnG
1 0.005575322 0.0005447632   0.005575322  2.674533e-46          0
```

CIT is based on testing the causality of three associations [96]. The analysis shows significant evidence for the causal path $LOY \to Gene \to Cancer$ when the maximum of the P-values obtained in each test (p_TassocL, p_TassocGgvnL, p_GassocLgvnT) is statistically significant. Here, we observe that Robjectp_cit is 0.00557 which rejects the null hypothesis of no causality. Therefore, we can conclude that the relationship between LOY and cancer is mediated by the expression of $DDX3Y$.

Mediation analysis, by contrast, directly tests the mediating effect of an *omic* data (transcription) with respect to another (LOY) in its relation with the trait [165]. Mediation analyses are employed in genetics to understand a known relationship by exploring the process by which a genetic variability modifies the risk of the disease. For instance, likely biological mediators between variability of gene structure and susceptibility to complex diseases are

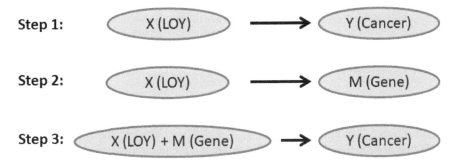

FIGURE 11.4
General mediation analysis schema adapted to the particular example of LOY
and cancer example.

mRNA expressions [62, 175]. However, unless mediation analysis is based on
experimental design, it does not imply causality. To test causality instrumen-
tal variable methods, such as Mendelian randomization are more appropriate
[122].

Mediation analysis comprises three regression models (see figure 11.4) [7].
The first regression model $Y = \beta_0 + \beta_1 X + \epsilon$ tests whether there is a significant
effect of X on Y (step 1 in figure 11.4). If there is no relationship between X
and Y, there is nothing to mediate. However, under a strong justified scientific
hypothesis, we can still test the mediation between X and Y, even when their
association is no significant [138].

In our example, we test the association between LOY(X) and cancer (Y),
adjusting by age [36]

```
> mod <- glm(Cancer ~ LOY + age , data=lusc, family="binomial")
> summary(mod)

Call:
glm(formula = Cancer ~ LOY + age, family = "binomial", data = lusc)

Deviance Residuals:
     Min        1Q    Median        3Q       Max
-2.96848   0.09825   0.16292   0.38028   1.04300

Coefficients:
            Estimate Std. Error z value Pr(>|z|)
(Intercept) -2.48803    1.71489  -1.451  0.14682
LOY1         2.66259    1.05675   2.520  0.01175 *
age          0.07813    0.02989   2.614  0.00894 **
---
Signif. codes:  0 '***' 0.001 '**' 0.01 '*' 0.05 '.' 0.1 ' ' 1

(Dispersion parameter for binomial family taken to be 1)
```

```
    Null deviance: 100.754  on 236  degrees of freedom
Residual deviance:  80.958  on 234  degrees of freedom
AIC: 86.958

Number of Fisher Scoring iterations: 7
```

We observe that both, LOY and age are highly associated with cancer status, confirming established findings [64]. The second regression model is M on X (step 2 in figure 11.4), $M = \beta_0 + \beta_2 X + \epsilon$. We test the significance of β_2. If the association is non-significant then M can be discarded as a mediator. In our example, we regress the transcription of $DDX3Y$ (M) on LOY (X) adjusting by age

```
> mod.M <- glm(DDX3Y ~ LOY + age, data=lusc)
> summary(mod.M)

Call:
glm(formula = DDX3Y ~ LOY + age, data = lusc)

Deviance Residuals:
     Min        1Q    Median        3Q       Max
-2.36474  -0.38608   0.06512   0.40642   1.79055

Coefficients:
              Estimate Std. Error t value Pr(>|t|)
(Intercept)  7.345620   0.280189  26.217   <2e-16 ***
LOY1        -1.719217   0.091548 -18.779   <2e-16 ***
age         -0.002081   0.004364  -0.477    0.634
---
Signif. codes:  0 '***' 0.001 '**' 0.01 '*' 0.05 '.' 0.1 ' ' 1

(Dispersion parameter for gaussian family taken to be 0.4945943)

    Null deviance: 291.17  on 236  degrees of freedom
Residual deviance: 115.74  on 234  degrees of freedom
AIC: 510.71

Number of Fisher Scoring iterations: 2
```

We observe a highly significant association $(P < 2e - 16)$. The third, and final, regression model for the mediation analysis is the effect of X on Y adjusting by M (step 3 in figure 11.4), $Y = \beta_0 + \beta_4 X + \beta_3 M + \epsilon$. We then ask whether β_4 is non-significant or smaller than β_1. If the effect of X on Y is totally accounted for, by the adjustment of M, M fully mediates the relationship between X and Y (full mediation). If the effect of X on Y is still significant, but smaller in magnitude, then M is a partial mediator. We test this in our example

```
> mod.Y <- glm(Cancer ~ LOY + DDX3Y + age, data=lusc, family="binomial")
> summary(mod.Y)

Call:
```

```
glm(formula = Cancer ~ LOY + DDX3Y + age, family = "binomial",
    data = lusc)

Deviance Residuals:
     Min         1Q     Median         3Q        Max
-2.52247    0.04098    0.13268    0.34086    1.36278

Coefficients:
            Estimate Std. Error z value Pr(>|z|)
(Intercept) 13.24193    6.95071   1.905   0.0568 .
LOY1         0.01266    1.32564   0.010   0.9924
DDX3Y       -2.14936    0.90506  -2.375   0.0176 *
age          0.07876    0.03146   2.504   0.0123 *
---
Signif. codes:  0 '***' 0.001 '**' 0.01 '*' 0.05 '.' 0.1 ' ' 1

(Dispersion parameter for binomial family taken to be 1)

    Null deviance: 100.754  on 236  degrees of freedom
Residual deviance:  73.276  on 233  degrees of freedom
AIC: 81.276

Number of Fisher Scoring iterations: 8
```

We observe that the effect of LOY on cancer is not significant after adjusting for *DDX3Y* expression, showing that effect of LOY in cancer is mediated by *DDX3Y* down-regulation.

We can test whether the mediation effect is significantly different from zero. There are two approaches available in R: the Sobel test with the `sobel` function in *multilevel* package [140] and the use of bootstrapping with the `mediate` function in *mediation* package [155]. The function `sobel` only allows continuous variables and models cannot be adjusted by other covariates. As such, `mediate` is more appropriate for our example. `mediate` takes the `glm` estimation of the models $X \rightarrow M$ and $X+M \rightarrow Y$, the names of the treatment (LOY) and mediator (gene expression) variables. For bootstrapping testing we set `boot = TRUE` and `sims=1000`.

```
> library(mediation)
> set.seed(12345)
> res <- mediate(mod.M, mod.Y, treat='LOY', mediator='DDX3Y')
> summary(res)

Causal Mediation Analysis

Quasi-Bayesian Confidence Intervals

                          Estimate 95% CI Lower 95% CI Upper p-value
ACME (control)             0.13059      0.03616         0.23   0.024 *
ACME (treated)             0.15061      0.00724         0.42   0.024 *
ADE (control)             -0.01757     -0.29260         0.15   0.962
ADE (treated)              0.00245     -0.03875         0.05   0.962
Total Effect               0.13304      0.05218         0.22   0.010 **
Prop. Mediated (control)   0.99757      0.40876         1.51   0.030 *
```

```
Prop. Mediated (treated)  0.95311     0.05006        3.88   0.030 *
ACME (average)                0.14060     0.02611        0.30   0.024 *
ADE (average)                -0.00756    -0.15785        0.09   0.962
Prop. Mediated (average)  0.97534     0.24857        2.64   0.030 *
---
Signif. codes:  0 '***' 0.001 '**' 0.01 '*' 0.05 '.' 0.1 ' ' 1

Sample Size Used: 237


Simulations: 1000
```

ADE corresponds to direct effect, while ACME stands for the mediation effect. The results indicate that the observed effect of LOY on cancer is mediated by gene expression of *DDX3Y* ($P = 0.024$). Additionally, the 97.5% of the observed variability is meditated by *DDX3Y*.

11.4.2   Domain-knowledge approach

Large international projects have annotated functional and pathway information. These include the Encyclopedia of DNA Elements (ENCODE) [24], the genotype-tissue expression (GTEx) project [85], the Roadmap Epigenomics Project [11] or the Kyoto Encyclopedia of Genes and Genomes (KEGG) [68], among many others. In domain-knowledge analyses, one considers whether regions with significant *omic* associations are within pathways and/or overlap with functional units, such as transcription factor binding sites, hypermethylated or hypomethylated regions, DNase sensitivity and regulatory motifs. The methods described in chapter 10 are examples where the significant transcriptomic regions correspond to genes enriched in gene pathways. Figure 11.5 illustrates how domain-knowledge guided approaches can be used to integrate information from different *omic* datasets.

A domain-knowledge analysis for genomic data is, for instance, the selection of candidate SNPs for a trait from multiple public databases, such as the GWAS catalog. The selected SNPs are then annotated in functional databases of motifs, transcription factors or quantitative trait loci (eQTL). SNPs with significant associations to those features or that, simply overlap, are then correlated to the phenotype of interest. This approach is therefore not hypothesis-free, like massive univariate testing, since the selection of features is based on current knowledge and is delimited to known scientific mechanisms. The main advantage is that results do not require further interpretation. Whereas, its main drawback is renouncing to significant findings that are not supported by existing knowledge, and therefore to the unbiased nature of the *omic* data.

Annotation for domain-knowledge analysis can be done with the *haploR* package which annotates data from HaploReg (`https://pubs.broadinst itute.org/mammals/haploreg/haploreg.php`) and RegulomeDB (`http://www.regulomedb.org`), which are web-based tools that annotate eQTL,

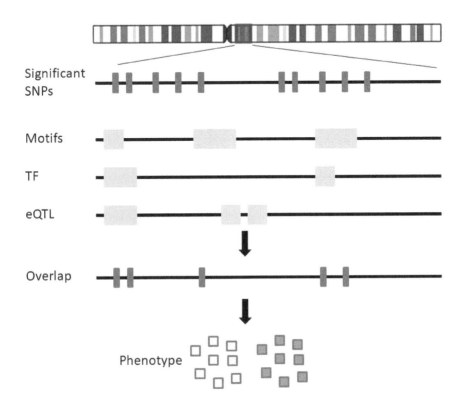

FIGURE 11.5

Domain-knowledge schema where a list of significant SNPs from any specific association study (genomic, transcriptomic, epigenomic) or obtained from public databases such as the GWAS catalog are first selected. Then, those candiate SNPs are annotated in funtional databases (motifs, transcription factors (TF), quantitative trait loci (eQTL). Only those SNPs which overlap in functional features are then tested with the phenotype of interest.

LD, motifs, etc., from large genomic projects such as ENCODE, the 1000 Genomes Project and the Roadmap Epigenomics Project, among others.

We illustrate the use of *haploR* with the results of the analysis of the GWAS for asthma, discussed in Chapter 4.2.3. We extract biological information such as eQTL, LD, motifs, etc., from the 2-top SNPs with the most significant associations

```
> library(haploR)
```

```
> x <- queryHaploreg(query=c("rs6931936", "rs12201995"))
> x
# A tibble: 20 x 35
      chr pos_hg38     r2  `D'` is_query_snp rsID   ref   alt     AFR   AMR
    <dbl> <chr>     <dbl> <dbl>        <dbl> <chr> <chr> <chr> <dbl> <dbl>
 1      6 45565552  0.99  1                0 rs10~ T     C      0.34  0.33
 2      6 45566644  0.99  1                0 rs12~ C     T      0.33  0.33
 3      6 45571327  1     1                1 rs12~ C     A      0.26  0.25
 4      6 45569192  0.98  1                0 rs12~ A     T      0.51  0.34
 5      6 45569194  0.98  1                0 rs12~ A     G      0.51  0.34
 6      6 45579965  0.87  0.93             0 rs13~ C     A      0.16  0.25
 7      6 45580867  0.81  0.9              0 rs13~ G     A      0.16  0.25
 8      6 45565416  1     1                0 rs16~ A     G      0.26  0.33
 9      6 45566082  0.87  1                0 rs19~ T     C      0.52  0.36
10      6 45564819  0.98  1                0 rs19~ C     A      0.26  0.32
11      6 45571599  0.98  0.99             0 rs35~ CAGG  C      0.32  0.25
12      6 45571683  1     1                0 rs35~ C     G      0.27  0.25
13      6 45569938  0.87  1                0 rs37~ T     G      0.51  0.36
14      6 45571409  0.98  1                0 rs68~ T     A      0.33  0.26
15      6 90029016  1     1                1 rs69~ A     G      0.49  0.42
16      6 45576627  0.88  0.99             0 rs69~ G     A      0.18  0.27
17      6 45567352  1     1                0 rs77~ G     A      0.26  0.33
18      6 45568622  0.99  1                0 rs77~ G     T      0.33  0.33
19      6 45568343  0.87  1                0 rs94~ T     A,C,G 0.52  0.36
20      6 45570576  0.96  1                0 rs94~ A     G      0.49  0.26
# ... with 25 more variables: ASN <dbl>, EUR <dbl>, GERP_cons <chr>,
#   SiPhy_cons <chr>, Chromatin_States <chr>,
#   Chromatin_States_Imputed <chr>, Chromatin_Marks <chr>, DNAse <chr>,
#   Proteins <chr>, eQTL <chr>, gwas <chr>, grasp <chr>, Motifs <chr>,
#   GENCODE_id <chr>, GENCODE_name <chr>, GENCODE_direction <chr>,
#   GENCODE_distance <chr>, RefSeq_id <chr>, RefSeq_name <chr>,
#   RefSeq_direction <chr>, RefSeq_distance <chr>,
#   dbSNP_functional_annotation <chr>, query_snp_rsid <chr>,
#   Promoter_histone_marks <fct>, Enhancer_histone_marks <fct>
```

The output shows the SNPs in high LD with the query SNPs, their frequency in 1000 genomes populations and reference alleles. We then search for motifs, GWAS, genes and histone marks associated with the query SNPs and those in high LD with them (> 0.9).

```
> sel <- as.numeric(x$r2) > 0.9
> x2 <- x[sel, c("pos_hg38", "rsID", "Motifs", "gwas",
+               "GENCODE_name", "Enhancer_histone_marks")]
> x2
```

```
# A tibble: 14 x 6
   pos_hg38 rsID    Motifs            gwas  GENCODE_name Enhancer_histone_ma~
   <chr>    <chr>   <chr>             <chr> <chr>        <fct>
 1 45565552 rs104~  Barx1;COMP1;Hox~  .     RUNX2        " "
 2 45566644 rs121~  E2A_2;E2A_5;Mrg~  .     RUNX2        ESDR, ESC, BRST, SK~
 3 45571327 rs122~  CTCF_disc9;Ik-1~  .     RUNX2        BLD, SKIN
 4 45569192 rs126~  BCL_disc4;HEY1_~  .     RUNX2        ESDR, BRST, BLD, SK~
 5 45569194 rs126~  .                 .     RUNX2        ESDR, BRST, BLD, SK~
 6 45565416 rs168~  Ik-2_3            .     RUNX2        " "
 7 45564819 rs192~  Pax-6_1;VDR_2     .     RUNX2        HRT, MUS
 8 45571599 rs350~  AP-1_disc4;GR_d~  .     RUNX2        BLD, SKIN
 9 45571683 rs356~  STAT_known5       .     RUNX2        BLD
10 45571409 rs680~  AP-1_disc9;TFII~  .     RUNX2        BLD, SKIN
11 90029016 rs693~  Hoxa4             .     BACH2        ESDR, BLD, BRN, THYM
12 45567352 rs775~  Dbx1;Foxp1;HMG-~  .     RUNX2        ESDR, ESC, BRST, BR~
13 45568622 rs775~  Mef2_disc3        .     RUNX2        ESDR, BRST, SKIN, H~
14 45570576 rs947~  Pou5f1_disc2      .     RUNX2        FAT
```

We can also specifically query RegulomeDB to identify DNA features and regulatory elements in non-coding regions of the human genome

```
> xx <- queryRegulome(c("rs6931936", "rs12201995"))
> xx$res.table
# A tibble: 2 x 5
  `#chromosome` coordinate rsid     hits                                 score
  <fct>         <fct>      <fct>    <fct>                                <fct>
1 chr6          90738734   rs69319~ Chromatin_Structure||DNase-seq|H~    5
2 chr6          45539063   rs12201~ Motifs|PWM||HIC2, Chromatin_Stru~    5
```

The column `score` provides the internal RegulomeDB score defined in http://www.regulomedb.org/help#score. SNPs with reported eQTLs effects can also be retrieved

```
> x3 <- x[, c("eQTL", "rsID")]
> x3
# A tibble: 20 x 2
   eQTL  rsID
   <chr> <chr>
 1 .     rs1041334
 2 .     rs12192817
 3 .     rs12201995
 4 .     rs12664796
 5 .     rs12664797
 6 .     rs13200687
 7 .     rs13205082
 8 .     rs16873510
 9 .     rs1928530
10 .     rs1928531
11 .     rs35068531
12 .     rs35611098
13 .     rs3763191
14 .     rs68057028
15 .     rs6931936
16 .     rs6938214
17 .     rs7752324
```

```
18 .      rs7758107
19 .      rs9463100
20 .      rs9472506
```

The Section 10.5 provides a formal test to verify whether any feature or genomic region overlaps with any region of interest or feature (motifs, promoters, enhancers, ...). The approach provides a P-value based on permutation testing that can be used to have statistical evidences of this method.

11.4.3 Meta-dimensional analysis

The second approach to integrating different *omic* datasets is data-driven. The meta-dimensional analysis includes multivariate methods to analyze multiple tables. Different methods can be applied to this end. For instance, multivariate analysis can be performed on the concatenation of tables, using the methods for single tables (e.g. PCA or MDS). Alternatively, one can also perform the analysis on each data type independently and then integrate the extracted variables in a multi-regression on the trait of interest.

11.4.3.1 Principal component analysis

Multivariate methods are commonly used to reduce the dimensionality of *omic* data and to provide descriptive visualization when multiple features are analyzed. Throughout the book, we have used these techniques to infer ancestry in genomic data or batch effects on transcriptomic and methylomic data.

Principal component analysis (PCA) is one of the multivariate methods that can be applied to any *omic* data X, of n individuals and m features $x = (x_1, x_2, \ldots, x_m)$. The PCA computes new orthogonal variables X such that they are linear combinations of the original variables $x_i = q_{i1}X_1 + q_{i2}X_2 + \ldots + q_{ip}X_p$ and their variances are decreasingly maximized. The coefficients of the linear combination q_{ij} are the loadings of subject i on the variable X_j and are ordered such that X_1 is the variable with the highest variance, X_2 with second highest variance, and so on. Figure 11.6 illustrates the transcription data of two genes across individuals. The principal components of the data are the projections of the subject data onto the axes of maximum decreasing variance.

Multivariate methods differ on the type of variables and the quantity to maximize. Principal co-ordinate analysis (PCoA), correspondence analysis (CA) and nonsymmetrical correspondence analysis (NSCA) can be used when variables are not normally distributed. PCoA, in particular, is applied when observations are binary or count data (e.g mutations or RNA-seq) while multi-dimensional scaling (MDS) is used when data are highly skewed. Independent component analysis (ICA), on the other hand, maximizes the statistical independence between the variables. In *omic* data, the number of features is larger than the number of subjects $m >> n$ and, therefore, single value decomposition is typically used to extract the first components. Other methods

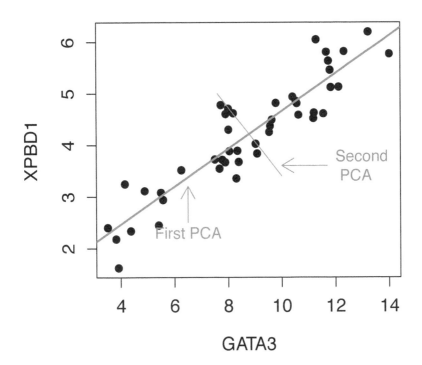

FIGURE 11.6
Hypothetical representation of gene expression of two genes on 45 indivuals.

based on a regularization step or L_1 penalization (Least Absolute Shrinkage and Selection Operator, LASSO) can also be applied. Sparse, penalized and regularized extensions of PCA and related methods have been used in the analysis of *omic* data [93].

We illustrate how to perform multivariate analysis using breast cancer data from TCGA described in Section 3.9. We start by performing a PCA on RNAseq data. There are several packages and functions in R for PCA. We particularly use *ade4*

```
> library(ade4)
> geneExpr <- breastMulti[["expression"]]
> dim(geneExpr)
[1] 8362    79
```

Features are expected in rows and subjects in columns. We compare the first two PCs (nf=2) using the function dudi.pca

```
> breastPCA <- dudi.pca(assay(geneExpr),
+                       scannf=FALSE, nf=2)
> summary(breastPCA)
Class: pca dudi
Call: dudi.pca(df = assay(geneExpr), scannf = FALSE, nf = 2)

Total inertia: 79

Eigenvalues:
    Ax1     Ax2     Ax3     Ax4     Ax5
 18.619   4.897   3.732   2.691   2.570

Projected inertia (%):
    Ax1     Ax2     Ax3     Ax4     Ax5
 23.568   6.199   4.724   3.406   3.253

Cumulative projected inertia (%):
    Ax1   Ax1:2   Ax1:3   Ax1:4   Ax1:5
  23.57   29.77   34.49   37.90   41.15

(Only 5 dimensions (out of 79) are shown)
```

We can see that the first PC explains 23.45% of the observed variability. The package *made4* can be used for visualizing the results

```
> library(made4)
> group <- geneExpr$ER.Status
> out <- ord(assay(geneExpr), type="pca", classvec=group)
>
> par(mfrow=c(2,1))
> cols <- c("black", "gray50")
> mycols <- ifelse(group=="Positive", cols[1], cols[2])
> plotarrays(out$ord$co, classvec=group, arraycol = cols)
> plotgenes(out, col="gray50")
```

The panels in figure 11.7 shows the projections of subjects and genes into

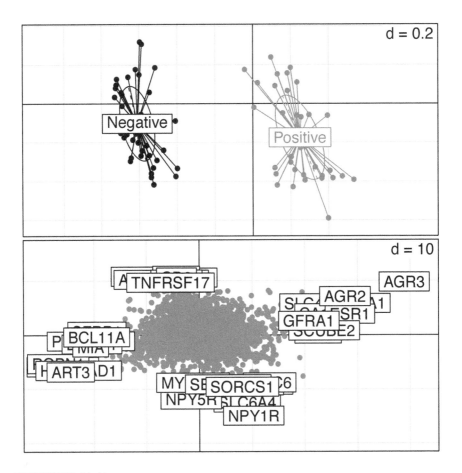

FIGURE 11.7
Gene expression principal component analysis of the BRCA dataset from the
TCGA project.

the principal axes. The variable group is used to color individuals and help interpret the results. We can see that the first axis clearly separates women with positive from negative ER. Therefore, genes located in the right-hand side of the first axes highly correlate with ER+ status while those in the left with correlate with ER-. The genes with the top highest loadings are obtained with

```
> ax1 <- topgenes(out, axis=1, n=5, ends="pos")
> ax2 <- topgenes(out, axis=2, n=5, ends="neg")
> cbind(pos=ax1, neg=ax2)
      pos      neg
[1,] "AGR3"  "NPY1R"
[2,] "FOXA1" "SLC6A4"
[3,] "ESR1"  "NPY5R"
[4,] "AGR2"  "SORCS1"
[5,] "FSIP1" "CST9L"
```

We can see that *AGR3*, *FOXA1*, *ESR1*, *AGR2*, *C1orf64* are the top-5 genes positively associated with ER+, while *NPY1R*, *SLC6A4*, *NPY5R*, *SORCS1*, *CST9L* are inversely associated with ER+ (i.e. are associated with ER-).

11.4.3.2   Sparse principal component analysis

Witten et al. [172] developed a penalized matrix decomposition for computing a rank-K approximation for a matrix based on L_1-penalization, implemented in the *PMA* package. This is a sparse version of PCA (sPCs) in which many loadings are expected to be small and, therefore, are penalized reducing them to zero. The input matrix expects individuals in rows and variables in columns. The argument sumabsv is used to define the level of sparseness in the data from 1 to $\sqrt{m}$, the lower the sparser; and the argument K defines the number of components to compute

```
> library(PMA)
> X <- t(assay(geneExpr))
> spca <- SPC(X, sumabsv=3, K=2)
```

sumabsv can be optimized with the function SPC.cv that provides the best value obtained from cross-validation error

```
> subabsvs.grid <- seq(10, sqrt(ncol(X)),len=20)
> spca.cv <- SPC.cv(X, sumabsvs=subabsvs.grid)
 Fold  1  out of  5
 Fold  2  out of  5
 Fold  3  out of  5
 Fold  4  out of  5
 Fold  5  out of  5
> spca.cv
Call:
SPC.cv(x = X, sumabsvs = subabsvs.grid)
```

```
Cross-validation errors:
   Sumabsvs CV Error CV S.E. # non-zero v's
1    10.000 141029.8 605.450          207.2
2    14.287 136746.1 597.392          454.6
3    18.573 132876.3 580.103          799.4
4    22.860 129393.3 567.546         1219.2
5    27.146 126265.6 558.658         1749.6
6    31.433 123508.9 548.901         2360.8
7    35.719 121103.5 548.020         3020.0
8    40.006 119022.1 545.749         3714.4
9    44.292 117276.4 543.262         4434.8
10   48.579 115883.4 541.978         5237.2
11   52.865 114836.7 540.258         6072.4
12   57.152 114123.4 534.497         6901.6
13   61.438 113740.1 529.189         7756.8
14   65.725 113666.8 527.246         8362.0
15   70.011 113666.8 527.246         8362.0
16   74.298 113666.8 527.246         8362.0
17   78.584 113666.8 527.246         8362.0
18   82.871 113666.8 527.246         8362.0
19   87.157 113666.8 527.246         8362.0
20   91.444 113666.8 527.246         8362.0

 Best sumabsv value (lowest CV error):  65.72482

 Smallest sumabsv value that has CV error within 1 SE of best CV error:  57.15177
```

Therefore, the SPC analysis with optimal `sumabsv` is 65.7248229 which can be used to perform sparse PCA as:

```
> spca <- SPC(X, sumabsv=spca.cv$bestsumabsv,
+             K=2)
```

We format the results

```
> rownames(spca$u) <- rownames(X)
> rownames(spca$v) <- colnames(X)
> head(spca$u)
                    [,1]         [,2]
TCGA-C8-A12V -0.1323889 -0.20192429
TCGA-A2-A0ST -0.1231891 -0.19372079
TCGA-E2-A159 -0.1101338 -0.12510040
TCGA-BH-A0BW -0.1478126 -0.14696896
TCGA-A2-A0SX -0.1310629 -0.13130637
TCGA-AR-A1AI -0.1848587 -0.07313812
> head(spca$v)
                    [,1]          [,2]
ABCA6   0.009946474 -0.0291634300
ABCC12  0.005426623  0.0071428801
ABCC5   0.002436294 -0.0039581021
ABCG8   0.002044495  0.0051294938
ABHD14B 0.009223434 -0.0007973728
ABR     0.004800245 -0.0020500351
```

and plot the individuals' loadings on the first two sparse PCs

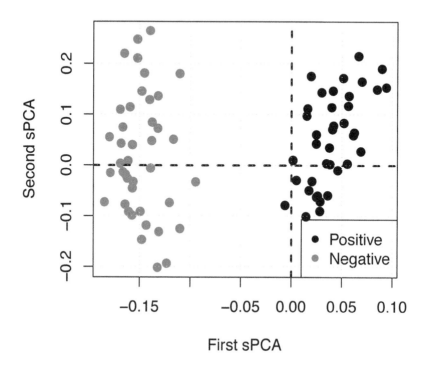

FIGURE 11.8
Gene expression sparse principal component analysis of the BRCA dataset
from the TCGA project.

```
> plot(spca$u, type="n", xlab="First sPCA",
+        ylab="Second sPCA")
> cols <- c("black", "gray50")
> mycols <- ifelse(group=="Positive", cols[1], cols[2])
> points(spca$u, pch=16, cex=1.3, col=mycols)
> legend("bottomright", c("Positive", "Negative"),
+        col=cols, pch=16)
> abline(h=0, lty=2, lwd=2)
> abline(v=0, lty=2, lwd=2)
> grid()
```

We see that the sPCs discriminate between women with positive ER from
those with negative ER. The top-5 genes associated to the first sPCs are

```
> ss <- spca$v[,1]
> ss.nonzero <- ss[ss!=0]
> ss.pos <- ss[order(ss, decreasing = TRUE)][1:5]
> ss.pos
     AGR3      FOXA1       ESR1       AGR2      FSIP1
0.10480527 0.08092237 0.07552895 0.07452137 0.06441611
```

The list of the top-5 genes agrees with the one obtained from the *made4* package. The object `ss.nonzero` contains all the genes that are informative for the first axis (e.g. are different from 0), which can be used to perform enrichment analyses. Note that here we do not test the association of individual genes with the phenotype but test how informative the features are in explaining the variability of the first component.

11.4.3.3 Canonical correlation and coinertia analyses

In our past examples, we have used multivariate methods to reduce the dimensionality of the data and study the extent the new variables can explain the phenotypic differences between individuals. When more than one *omic* data is available, we can perform a canonical correlation analysis (CCA) which is an extension of PCA for more than two tables x and y. CCA computes new variables $X = (X_1, ...X_m)$, $Y = (Y_1, ...Y_l)$ for each *omic* dataset such that they are linear combinations of the original variables

$$x = pX$$

$$y = qY,$$

where p and q are the loading matrices, and the correlations among X and Y are decreasingly maximized. X_i and Y_i are known as i-th canonical variables and their correlations are the canonical correlations. CCA has been used to analyze *omic* datasets [93]. Its main limitation is that the number of features greatly exceeds the number of subjects and hence parameter estimation cannot be applied using standard methods. Additionally, most variables have little or no weight in the canonical axes. Therefore, penalized (sparse) CCA has been proposed [173]. The method adds a penalty term, that is a multiple of the identity, by the scalar λ, to the correlation between the canonical variables. Therefore the method aims to decreasingly maximize $cor(X_i, Y_i + \lambda I)$, for $i = 1..min(m, l)$.

Coinertia analysis (CIA) differs from CCA in that it decreasingly maximizes the covariance between the variables $cov^2(X_i, Y_i)$ rather than their correlation. CIA does not require an inversion step of the covariance matrix; thus, regularization or penalization implementation is not needed. As covariance is generally defined for different types of variables, CIA can be used for quantitative or qualitative variables, in contrast to CCA that is a method for continuous variables only. In CIA one can compute the coefficient RV that measures the similarity between datasets with values between 0 and 1. The closer it is to 1 the greater the global similarity between the two datasets is [92].

We demonstrate CCA and CIA on the multi-*omic* dataset of breast cancer. We analyze transcriptomic and proteomic data. We apply the sparse version of CCA implemented in *PMA* package. We start by calling the function `MultiCCA.permute` to optimize the penalty and compute the initialization parameters for the sparse CCA, such as `ws.init`. The argument `type` indicates whether the features of each table are ordered or not. When data are ordered (e.g. by chromosome and genomic position) a fused lasso penalty is applied to enforce both sparsity and smoothness. We compute sparse CCA with the function RMultiCCA

```
> library(PMA)
> df1 <- t(assay(breastMulti[["expression"]]))
> df2 <- t(exprs(breastMulti[["RPPA"]]))
>
> ddlist <- list(df1, df2)
> perm.out <- MultiCCA.permute(ddlist,
+                              type=c("standard", "standard"),
+                              trace=FALSE)
> resMultiCCA <- MultiCCA(ddlist, ncomponents=1,
+                         penalty=perm.out$bestpenalties,
+                         ws=perm.out$ws.init,
+                         type=c("standard", "standard"),
+                         trace=FALSE)
```

We inspect the results

```
> rownames(resMultiCCA$ws[[1]]) <- colnames(df1)
> rownames(resMultiCCA$ws[[2]]) <- colnames(df2)
> head(resMultiCCA$ws[[1]])
                [,1]
ABCA6   0.001561697
ABCC12  0.003019194
ABCC5   0.000000000
ABCG8   0.000000000
ABHD14B 0.021655906
ABR     0.000000000
> head(resMultiCCA$ws[[2]])
                     [,1]
c.Myc       -0.05343571
HER3         0.00000000
XBP1         0.00000000
Fibronectin  0.00000000
PAI.1        0.00000000
p21          0.00000000
```

We are interested in selecting those features of both datasets with loadings that are different from 0 in the first axis, and associate with differences between positive and negative ER. The list of features for the first table (genes) can be obtained by

```
> genes <- resMultiCCA$ws[[1]]
> genes.sig <- genes[genes[,1]!=0,]
> head(genes.sig)
        ABCA6          ABCC12        ABHD14B         ACOT2          ACSL1
 0.001561697   0.003019194   0.021655906   0.024808183  -0.003452103
          AGPS
-0.002235587
> length(genes.sig)
[1] 4984
```

and for proteins

```
> proteins <- resMultiCCA$ws[[2]]
> proteins.sig <- proteins[proteins[,1]!=0,]
> head(proteins.sig)
      c.Myc     Annexin_I     PKC.alpha     YAP_pS127        COX.2         Chk1
-0.05343571  -0.04143280  -0.01122895   0.01257553  -0.04544412  -0.10109053
> length(proteins.sig)
[1] 105
```

The result is a list of genes and proteins whose correlation contributes to the first sCCA axis, which discriminates ER+ and ER- status. We thus have a list of features that are considered from two different *omic* datasets and whose correlation contribute to discriminate phenotypic differences between individuals. Note that `multiCCA` can analyze more than two *omic* datasets, adding more tables into the `ddlist` object.

Coinertia analysis can be applied with the *made4* and *omicade4 made4* Bioconductor packages. We perform a CIA on our previous example using the function `cia`. The function requires two arguments one for each *omic* datasets (matrix, data frame or ExpressionSet), and expects individuals in columns and variables in rows

```
> library(made4)
> library(omicade4)
> resCIA <- cia(assay(breastMulti[["expression"]]),
+               exprs(breastMulti[["RPPA"]]))
```

Figure 11.9 shows the projection of samples (top plot) and features (bottom plots) on the two principal axes. We observe that the first axis discriminates again women by ER status.

```
> mycols <- c("black", "gray50")
> cols <- ifelse(group=="Positive",
+                mycols[1], mycols[2])
> plot(resCIA, classvec=group, nlab=3, clab=0,
+      cpoint=3, col=cols)
```

The top-5 features that most positively contribute to the first axis can be retrieved by

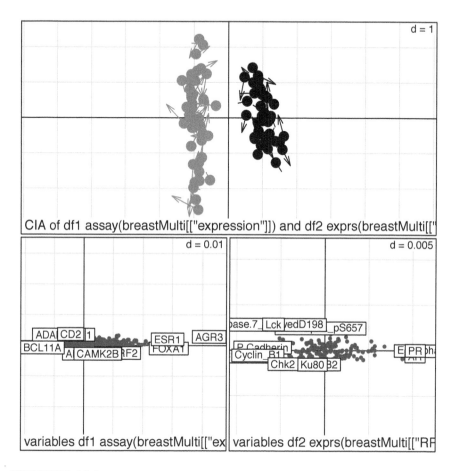

FIGURE 11.9
Gene expression and protein coinertia analysis of the BRCA dataset from the
TCGA project.

```
> topVar(resCIA, axis=1, topN=5, end="positive")
  ax1_df1_positive ax1_df2_positive
1             AGR3         ER.alpha
2            FOXA1               PR
3             ESR1               AR
4             AGR2           INPP4B
5            FSIP1            GATA3
```

while the those that most negatively contribute to the first axis are

```
> topVar(resCIA, axis=1, topN=5, end="negative")
  ax1_df1_negative     ax1_df2_negative
1           ROPN1            Cyclin_E1
2          ROPN1B            Cyclin_B1
3          BCL11A            P.Cadherin
4            ART3 Caspase.7_cleavedD198
5           SFRP1                 MSH6
```

CIA analysis can also be applied to more than two tables. For multiple *omic* data, CIA can be done with the function `mcia`. We apply it in our example, for six different *omic* data. The function requires a list of tables. *MultiDataSet* contains a function (`as.list`) to coerce the *omic* datasets into a list. We illustrate the analysis of gene expression, miRNA, RPPA and methylation datasets

```
> ll <- as.list(breastMulti)
> names(ll)
[1] "methylation"   "expression"    "miRNA"        "miRNAprecursor"
[5] "RPPA"          "LOH"           "CNA"
> resMCIA <- mcia(ll[ c("methylation", "expression",
+                       "miRNA", "RPPA") ] )
```

Figure 11.10 shows the graphic representation of multiple coinertia analysis.

```
> mycols <- c("black", "gray50")
> cols <- ifelse(group=="Positive", mycols[1], mycols[2])
> plot(resMCIA, axes=1:2, sample.lab=FALSE,
+      sample.legend=FALSE, phenovec=group,
+      gene.nlab=2, sample.col=cols, df.pch=2:5,
+      df.color=c("black", "gray30", "gray50", "gray90"))
```

The top-left panel shows once again that individuals are well clustered along the first axis by their ER status. The top-right panel shows the projection into the first two CIA axes of features across all *omic* datasets, highlighting the features that are more correlated with each axis. The bottom-left panel provides the eigen-values from the analysis that are used to measure the amount of variability explained by each axis. We observe, for instance, that the first axis explains a large amount of variance. The bottom-right plot shows the RV coefficient that measures the similarity among the *omic* datasets. We see that proteins and gene expression are the most similar datasets. RV also

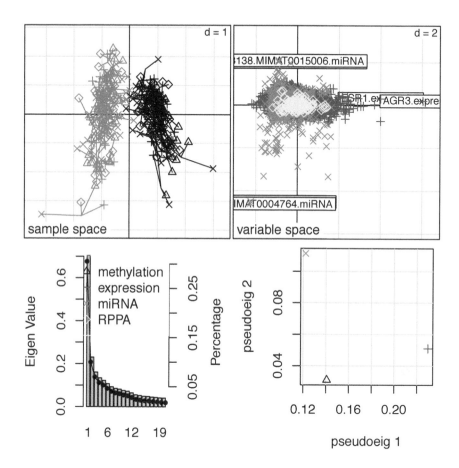

FIGURE 11.10
Multiple coinertia analysis of the BRCA dataset from the TCGA project.

informs on the *omic* datasets that are more important for each axis. We see that proteins and gene expression are highly correlated with the first axis while the second axis is mainly driven by miRNA.

We can retrieve the top-5 features that have the highest weight on the positive direction on axis 1

```
> topVar(resMCIA, end="positive", axis=1, topN=5)
  ax1_methylation_positive ax1_expression_positive
1     cg09952204.methylation          AGR3.expression
2     cg08097882.methylation          ESR1.expression
3     cg04988423.methylation          FOXA1.expression
4     cg00679738.methylation          AGR2.expression
5     cg12601757.methylation          FSIP1.expression
                       ax1_miRNA_positive ax1_RPPA_positive
1 hsa.mir.4254.MIMAT0016884.miRNA       ER.alpha.RPPA
2 hsa.mir.302b.MIMAT0000715.miRNA              AR.RPPA
3 hsa.mir.1197.MIMAT0005955.miRNA              PR.RPPA
4 hsa.mir.3945.MIMAT0018361.miRNA          INPP4B.RPPA
5  hsa.mir.198.MIMAT0000228.miRNA           GATA3.RPPA
```

11.4.3.4 Regularized generalized canonical correlation

The previous methods assume no intrinsic ordering on the *omic* data. In particular, the methods estimate the loadings by maximizing the correlation or covariance among tables with no causal relationships among them. However, one may assume an ordering motivated from directional mechanisms, such as genomics variables affect transcriptomics variables that in turn affect proteomic features. Directional models are implemented by path modeling and structural equation models (SEM) [161]. Regularized Generalized Canonical Correlation (RGCCA) provides a unified framework for directional modeling for multivariate integrative approaches [153]. Additionally, RGCCA integrates a feature selection method, named sparse GCCA (SGCCA) that can be used to select informative features for downstream analyses [152].

The objective of RGCCA is to compute, for each *omic* data, weighted composite variables (called block components) $\mathbf{y}_j = \mathbf{X}_j\mathbf{a}_j, j = 1, \ldots, J$ (where $\mathbf{a}_j$ is a column-vector with p_j elements) summarizing the relevant information between and within the blocks. The block components are obtained such that they explain well their own block and they are highly correlated with the other components that are assumed to be mechanistically connected.

Details about RGCCA mathematical definition and estimation can be found in [153, 154]. In summary, specific models and multivariate methods (PCA, CCA, CIA) can be defined through different parameters. One of those parameters is the family function g that selects the type of multivariate analysis, we want to perform. For instance, when g is the identity, maximization is performed on the sum of covariances between block components; when g is the absolute value, maximization is done on the sum of the absolute values of the covariances; the square function that maximizes the sum of squared covariances.

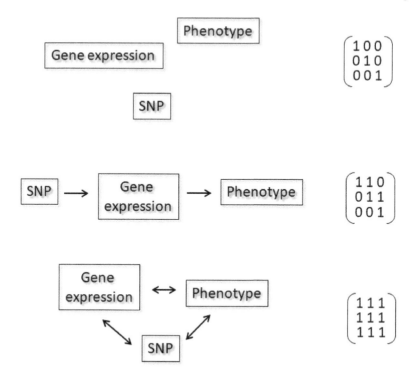

FIGURE 11.11
Different connections between genomic, transcriptomic and phenotypic data
that can be assessed using RGCCA by changing the design matrix.

The parameter m defines the relative contribution of each *omic* dataset. A
fair model is a model where all blocks contribute equally to the solution and
thus $m = 1$. However, $m > 1$ is preferable for discriminating between blocks.
In practice, m is equal to 1, 2, 3 or 4. The higher the value of m the more the
method acts as block selector.

The design matrix $\mathbf{C}$ is a symmetric $J \times J$ matrix of nonnegative elements describing the network of connections between blocks to be expected
from likely biological mechanisms. Usually, $c_{jk} = 1$ for two connected blocks
and 0 otherwise. Figure 11.11 shows different connections that we assume in
the analysis of genomic, transcriptomic and phenotypic data. The top figure
represents data that are not connected across modalities. In the second example, gene expression is mediating the association between SNPs and the
phenotype. In the third example, all data is correlated across modalities.

The τ_j are shrinkage parameters ranging from 0 to 1 and interpolates
smoothly between maximizing the correlation and maximizing the covariance.
The parameter selects intermediate multivariate methods continuously rang-

ing from CCA or CIA. CCA is obtained when $\tau_j = 0$, which can lead to unstable solutions in case of multi-collinearity and cannot be used when a data block is rank deficient (e.g. $n < p_j$). CIA is obtained when $\tau_j = 1$ and it leads to stable components (large variance). Variance dominates over correlation, thus $0 < \tau_j < 1$ is a compromise between stability and correlation. This setting can be used for an *omic* dataset that is rank deficient. That is when a sparse or penalized approach is required.

We demonstrate the use of RGCCA with the `rgcca` function of the *RGCCA* package. In particular, we show how to perform different multivariate methods on our breast cancer example. Within *RGCCA*, PCA is performed as follows

```
> library(RGCCA)
>
> X <- t(assay(breastMulti[["expression"]]))
>
> # one omic X
> # Design matrix C
> # Shrinkage parameters tau = c(tau1, tau2)
> pca.with.rgcca <- rgcca(A = list(X, X),
+                         C = matrix(c(0, 1, 1, 0), 2, 2),
+                         tau = c(1, 1),
+                         ncomp = c(2, 2))
Computation of the RGCCA block components based on the centroid scheme
Shrinkage intensity paramaters are chosen manually
Computation of the RGCCA block components #1 is under progress...
 Iter:    1 Fit: 0.35686962 Dif:  -0.00000000
The RGCCA algorithm converged to a stationary point after 0 iterations
Computation of the RGCCA block components #2 is under progress ...
 Iter:    1 Fit: 0.15228006 Dif:  -0.00000000
The RGCCA algorithm converged to a stationary point after 0 iterations
```

Figure 11.12 confirms the PCA results previously obtained for RNA-seq data. *RGCCA* does not contain specific functions to visualize the results. We have created functions to plot individual and variable projections into principal components, as obtained by *RGCCA*, as well as functions to extract the top variables of each axis. The functions `plotInd` and `topVar` are available on **https://github.com/isglobal-brge/book_omic_association/tree/master/R**.

Figure 11.12 shows the result of applying the functions on the results in `pca.with.rgcca`

```
> plotInd(pca.with.rgcca, group=group,
+         col.list=c("gray50", "black"))
```

`rgcca` can also be used to perform a CCA for the integration of the RNA-seq and miRNA data of our example. CCA is called as follows

```
> X <- t(assay(breastMulti[["expression"]]))
> Y <- t(exprs(breastMulti[["miRNA"]]))
> cca.with.rgcca <- rgcca(A= list(X, Y),
+                         C = matrix(c(0, 1, 1, 0), 2, 2),
```

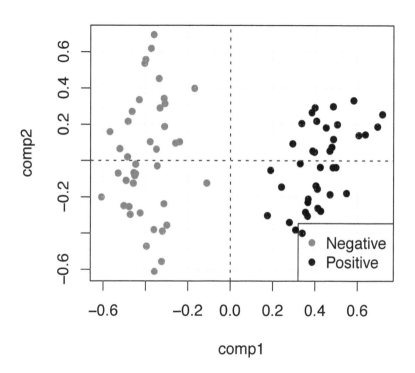

FIGURE 11.12
Principal component analysis using RGCCA on the BRCA dataset from the
TCGA project.

```
+                              tau = c(0, 0),
+                              ncomp=c(2,2))
Computation of the RGCCA block components based on the centroid scheme
Shrinkage intensity paramaters are chosen manually
Computation of the RGCCA block components #1 is under progress...
 Iter:    1  Fit: 2.00000000  Dif:  0.36285653
 Iter:    2  Fit: 2.00000000  Dif:  0.00000000
The RGCCA algorithm converged to a stationary point after 1 iterations
Computation of the RGCCA block components #2 is under progress ...
 Iter:    1  Fit: 2.00000000  Dif:  1.56213486
 Iter:    2  Fit: 2.00000000  Dif:  0.00000000
The RGCCA algorithm converged to a stationary point after 1 iterations
```

The following code illustrates non-executable syntax to call `rgcca` for a Generalized Canonical Correlation Analysis (GCCA) on J *omic* data sets

```
> # X1 = omic1, ..., XJ = omicJ, X_{J+1} = [X1, ..., XJ]
> # (J+1)*(J+1) Design matrix C
> C = matrix(c(0, 0, 0, ..., 0, 1,
+              0, 0, 0, ..., 0, 1,
+              ...,
+              1, 1, 1, ..., 1, 0), J+1, J+1)
> #Shrinkage parameters tau = c(tau1, ...,  tauJ, tau_{J+1})
> gcca.with.rgcca = rgcca(A= list(X1, ..., XJ, cbind(X1, ..., XJ)),
+                      C = C, tau = rep(0, J+1),
+                      scheme = "factorial")
```

and for MCIA

```
> # X1 = omic1, ..., XJ = omicJ, X_{J+1} = [X1, ..., XJ]
> # (J+1)*(J+1) Design matrix C
> C = matrix(c(0, 0, 0, ..., 0, 1,
+              0, 0, 0, ..., 0, 1,
+              ...,
+              1, 1, 1, ..., 1, 0), J+1, J+1)
> # Shrinkage parameters tau = c(tau1, ...,  tauJ, tau_{J+1})
> mcia.with.rgcca = rgcca(A= list(X1, ..., XJ, cbind(X1, ..., XJ)),
+                      C = C, tau = c(rep(1, J), 0),
+                      scheme = "factorial")
```

Turning back to our breast cancer example, we apply the previous syntax in the analysis of miRNA and RNA-seq and protein levels, where we assume a correlation pathway between the three different *omic* datasets. We first standardize the data to have zero mean and unit variance, to ensure comparability across variable units.

```
> X <- t(assay(breastMulti[["expression"]]))
> Y <- t(exprs(breastMulti[["miRNA"]]))
> Z <- t(exprs(breastMulti[["RPPA"]]))
> A <- list(X,Y,Z)
> A <- lapply(A, function(x) scale2(x, bias = TRUE))
```

The `rgcca` function allows standardization by setting the `scale` argument to TRUE. Within the function, the variables in an *omic* dataset are standardized and then divided by the square root of the number of variables in the *omic* dataset.

We model the case where miRNA and RNA-seq affect protein levels using the **C** matrix

```
> C <- matrix(c(1,0,1,0,1,1,0,0,1), nrow=3, byrow = TRUE)
> C
     [,1] [,2] [,3]
[1,]    1    0    1
[2,]    0    1    1
[3,]    0    0    1
```

We use the factorial scheme $(g(x) = x^2)$ and mode B for all blocks corresponding to the full correlation criterion

```
> rgcca.factorial <- rgcca(A, C=C, tau = rep(0, 3),
+                          scheme ="factorial",
+                          ncomp=c(2,2,2), scale = FALSE,
+                          verbose = FALSE)
```

The weight vectors, solution of the optimization problem, are obtained as:

```
> head(rgcca.factorial$a[[1]]) #for RNAseq
                 [,1]           [,2]
ABCA6    -5.806754e-05 -0.0007684196
ABCC12    1.159874e-04  0.0003331290
ABCC5    -1.565170e-04  0.0008968399
ABCG8     5.707533e-04  0.0013436526
ABHD14B  -2.098797e-05 -0.0012613653
ABR      -1.271363e-04  0.0000369512
> head(rgcca.factorial$a[[2]]) #for miRNA
                                [,1]          [,2]
hsa-let-7a-2.MIMAT0010195  0.010486366  0.004498749
hsa-let-7a.MIMAT0000062    0.011374213  0.003922677
hsa-let-7a.MIMAT0004481   -0.001822565 -0.011102683
hsa-let-7b.MIMAT0000063    0.011090914 -0.011768135
hsa-let-7b.MIMAT0004482    0.013784704 -0.013143639
hsa-let-7c.MIMAT0000064   -0.005595051  0.018529147
> head(rgcca.factorial$a[[3]]) #for proteins
                 [,1]          [,2]
c.Myc       -0.0091590758 -0.018479311
HER3         0.0078303648 -0.001808814
XBP1         0.0014584756 -0.006494360
Fibronectin -0.0004110668  0.022852916
PAI.1       -0.0085921031  0.031467929
p21          0.0025084042  0.021615805
```

The plot of individuals projected into the first two axes are shown in Figure 11.13.

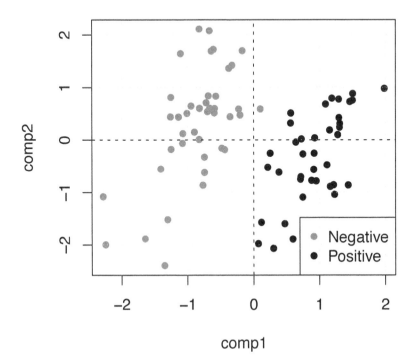

FIGURE 11.13
Multiomic data analysis using RGCCA on the BRCA dataset from the TCGA project.

```
> plotInd(rgcca.factorial, group=group,
+          col.list=c("gray50", "black"))
```

We observe again that the first axis discriminates women by ER status. The top features that are possitively associated with axis 1, that is, those associated with ER+, can be obtained with function `topVars`

```
> topVars(rgcca.factorial, axis=1, end="pos", topN=5)
        table_1    table_2                        table_3
top_1 "PGLYRP2" "hsa-mir-29b-2.MIMAT0004515" "ER.alpha"
top_2 "TMEM163" "hsa-mir-29c.MIMAT0004673"   "PR"
top_3 "CYP4F11" "hsa-mir-190b.MIMAT0004929"  "GATA3"
top_4 "FDXR"    "hsa-mir-664.MIMAT0005949"   "AR"
top_5 "PFN3"    "hsa-mir-342.MIMAT0004694"   "INPP4B"
```

The top features associated with ER- are obtained by

```
> topVars(rgcca.factorial, axis=1, end="neg", topN=5)
      table_1  table_2                        table_3
top_1 "COL9A2" "hsa-mir-548a-3.MIMAT0004803"  "Cyclin_E1"
top_2 "SIX4"   "hsa-mir-155.MIMAT0000646"     "Cyclin_B1"
top_3 "RASD1"  "hsa-mir-1247.MIMAT0005899"    "P.Cadherin"
top_4 "GNE"    "hsa-mir-939.MIMAT0004982"     "MSH6"
top_5 "IFNA17" "hsa-mir-1275.MIMAT0005929"    "Rad51"
```

RGCCA computes the Average Variance Explained (AVE) for three different cases: For all the block, for the inner model, and for each block. The former one is denoted $AVE(\mathbf{X}_j)$, varies between 0 and 1 and reflects the proportion of variance captured by $\mathbf{y}_j$. In the results object `rgcca.factorial`, these metrics are stored in the field `AVE`

```
> rgcca.factorial$AVE
$AVE_X
$AVE_X[[1]]
     comp1       comp2
0.14672906 0.03034896

$AVE_X[[2]]
     comp1       comp2
0.06683530 0.02441686

$AVE_X[[3]]
     comp1       comp2
0.17983101 0.07924063


$AVE_outer_model
[1] 0.13948314 0.03065012

$AVE_inner_model
[1] 1.0000000 0.9999999
```

We can see, for instance, that the AVE of RNA-seq (first table) by the first axis is 14.7% while the first axis explains the 18% of variability observed in proteins.

Bibliography

[1] Adrian Alexa, Jörg Rahnenführer, and Thomas Lengauer. Improved scoring of functional groups from gene expression data by decorrelating go graph structure. *Bioinformatics*, 22(13):1600–1607, 2006.

[2] Simon Anders and Wolfgang Huber. Differential expression analysis for sequence count data. *Genome Biology*, 11(10):R106, 2010.

[3] Carl A Anderson, Fredrik H Pettersson, Geraldine M Clarke, Lon R Cardon, Andrew P Morris, and Krina T Zondervan. Data quality control in genetic case-control association studies. *Nature Protocols*, 5(9):1564, 2010.

[4] Martin J Aryee, Andrew E Jaffe, Hector Corrada-Bravo, Christine Ladd-Acosta, Andrew P Feinberg, Kasper D Hansen, and Rafael A Irizarry. Minfi: A flexible and comprehensive Bioconductor package for the analysis of infinium DNA methylation microarrays. *Bioinformatics*, 30(10):1363–1369, 2014.

[5] Philip Asherson, Keeley Brookes, Barbara Franke, Wai Chen, Michael Gill, Richard P Ebstein, Jan Buitelaar, Tobias Banaschewski, Edmund Sonuga-Barke, Jacques Eisenberg, et al. Confirmation that a specific haplotype of the dopamine transporter gene is associated with combined-type ADHD. *American Journal of Psychiatry*, 164(4):674–677, 2007.

[6] David J Balding. A tutorial on statistical methods for population association studies. *Nature Reviews Genetics*, 7(10):781, 2006.

[7] Reuben M Baron and David A Kenny. The moderator–mediator variable distinction in social psychological research: Conceptual, strategic, and statistical considerations. *Journal of Personality and Social Psychology*, 51(6):1173, 1986.

[8] Tanya Barrett, Stephen E Wilhite, Pierre Ledoux, Carlos Evangelista, Irene F Kim, Maxim Tomashevsky, Kimberly A Marshall, Katherine H Phillippy, Patti M Sherman, Michelle Holko, et al. NCBI GEO: Archive for functional genomics data sets update. *Nucleic Acids Research*, 41(D1):D991–D995, 2012.

[9] Ferdouse Begum, Debashis Ghosh, George C Tseng, and Eleanor Feingold. Comprehensive literature review and statistical considerations for GWAS meta-analysis. *Nucleic Acids Research*, 40(9):3777–3784, 2012.

[10] Y. Benjamini and Y. Hochberg. Controlling the false discovery rate: A practical and powerful approach to multiple testing. *Journal of the Royal Statistical Societ Series B*, 57:289–300, 1995.

[11] Bradley E Bernstein, John A Stamatoyannopoulos, Joseph F Costello, Bing Ren, Aleksandar Milosavljevic, Alexander Meissner, Manolis Kellis, Marco A Marra, Arthur L Beaudet, Joseph R Ecker, et al. The nih roadmap epigenomics mapping consortium. *Nature Biotechnology*, 28(10):1045, 2010.

[12] Richard Bourgon, Robert Gentleman, and Wolfgang Huber. Independent filtering increases detection power for high-throughput experiments. *Proceedings of the National Academy of Sciences*, 107(21):9546–9551, 2010.

[13] James H Bullard, Elizabeth Purdom, Kasper D Hansen, and Sandrine Dudoit. Evaluation of statistical methods for normalization and differential expression in mrna-seq experiments. *BMC Bioinformatics*, 11(1):94, 2010.

[14] S van Buuren and Karin Groothuis-Oudshoorn. mice: Multivariate imputation by chained equations in R. *Journal of Statistical Software*, pages 1–68, 2010.

[15] Ruiz C, Caceres A, Lopez-Sanchez M, Tolosana I, Perez-Jurado L, and Gonzalez JR. scoreinvhap: Inversion genotyping for genome-wide association studies. *Plos Genetics*, 2019.

[16] Alejandro Cáceres and Juan R González. Following the footprints of polymorphic inversions on SNP data: From detection to association tests. *Nucleic Acids Research*, 43(8):e53–e53, 2015.

[17] Alejandro Cáceres, Suzanne S Sindi, Benjamin J Raphael, Mario Cáceres, and Juan R González. Identification of polymorphic inversions from genotypes. *BMC Bioinformatics*, 13(1):28, 2012.

[18] Ian M Campbell, Chad A Shaw, Pawel Stankiewicz, and James R Lupski. Somatic mosaicism: Implications for disease and transmission genetics. *Trends in Genetics*, 31(7):382–392, 2015.

[19] Marc Carlson and Bioconductor Package Maintainer. *TxDb.Hsapiens.UCSC.hg19.knownGene: Annotation package for TxDb object(s)*, 2015. R package version 3.2.2.

[20] Andrew G Clark. Inference of haplotypes from PCR-amplified samples of diploid populations. *Molecular Biology and Evolution*, 7(2):111–122, 1990.

[21] David G Clayton, Neil M Walker, Deborah J Smyth, Rebecca Pask, Jason D Cooper, Lisa M Maier, Luc J Smink, Alex C Lam, Nigel R Ovington, Helen E Stevens, et al. Population structure, differential bias and genomic control in a large-scale, case-control association study. *Nature Genetics*, 37(11):1243, 2005.

[22] Leonardo Collado-Torres, Abhinav Nellore, Kai Kammers, Shannon E Ellis, Margaret A Taub, Kasper D Hansen, Andrew E Jaffe, Ben Langmead, and Jeffrey T Leek. Reproducible RNA-seq analysis using recount2. *Nature Biotechnology*, 35(4):319–321, 2017.

[23] 1000 Genomes Project Consortium et al. A global reference for human genetic variation. *Nature*, 526(7571):68–74, 2015.

[24] ENCODE Project Consortium et al. The encode (encyclopedia of DNA elements) project. *Science*, 306(5696):636–640, 2004.

[25] International HapMap Consortium et al. The international hapmap project. *Nature*, 426(6968):789, 2003.

[26] International Parkinson Disease Genomics Consortium et al. Imputation of sequence variants for identification of genetic risks for Parkinson's disease: A meta-analysis of genome-wide association studies. *The Lancet*, 377(9766):641–649, 2011.

[27] Wellcome Trust Case Control Consortium et al. Genome-wide association study of 14,000 cases of seven common diseases and 3,000 shared controls. *Nature*, 447(7145):661, 2007.

[28] EH Corder, AM Saunders, WJ Strittmatter, DE Schmechel, PC Gaskell, GWet al Small, AD Roses, JL Haines, and M Al Pericak-Vance. Gene dose of apolipoprotein e type 4 allele and the risk of Alzheimers disease in late onset families. *Science*, 261(5123):921–923, 1993.

[29] Christina Curtis, Sohrab P Shah, Suet-Feung Chin, Gulisa Turashvili, Oscar M Rueda, Mark J Dunning, Doug Speed, Andy G Lynch, Shamith Samarajiwa, Yinyin Yuan, et al. The genomic and transcriptomic architecture of 2,000 breast tumours reveals novel subgroups. *Nature*, 486(7403):346–352, 2012.

[30] Pan Du, Warren A Kibbe, and Simon M Lin. lumi: A pipeline for processing illumina microarray. *Bioinformatics*, 24(13):1547–1548, 2008.

[31] Frank Dudbridge. Power and predictive accuracy of polygenic risk scores. *PLoS Genetics*, 9(3):e1003348, 2013.

[32] Jan P Dumanski, Jean-Charles Lambert, Chiara Rasi, Vilmantas Giedraitis, Hanna Davies, Benjamin Grenier-Boley, Cecilia M Lindgren, Dominique Campion, Carole Dufouil, Florence Pasquier, et al. Mosaic loss of chromosome y in blood is associated with Alzheimer's disease. *The American Journal of Human Genetics*, 98(6):1208–1219, 2016.

[33] Georg B Ehret, Patricia B Munroe, Kenneth M Rice, Murielle Bochud, Andrew D Johnson, Daniel I Chasman, Albert V Smith, Martin D Tobin, Germaine C Verwoert, Shih-Jen Hwang, et al. Genetic variants in novel pathways influence blood pressure and cardiovascular disease risk. *Nature*, 478(7367):103, 2011.

[34] Laurent Excoffier and Montgomery Slatkin. Maximum-likelihood estimation of molecular haplotype frequencies in a diploid population. *Molecular biology and evolution*, 12(5):921–927, 1995.

[35] Valery L Feigin, Gregory A Roth, Mohsen Naghavi, Priya Parmar, Rita Krishnamurthi, Sumeet Chugh, George A Mensah, Bo Norrving, Ivy Shiue, Marie Ng, et al. Global burden of stroke and risk factors in 188 countries, during 1990–2013: A systematic analysis for the global burden of disease study 2013. *The Lancet Neurology*, 15(9):913–924, 2016.

[36] Lars A Forsberg, Chiara Rasi, Niklas Malmqvist, Hanna Davies, Saichand Pasupulati, Geeta Pakalapati, Johanna Sandgren, Teresita Diaz de Ståhl, Ammar Zaghlool, Vilmantas Giedraitis, et al. Mosaic loss of chromosome y in peripheral blood is associated with shorter survival and higher risk of cancer. *Nature Genetics*, 46(6):624, 2014.

[37] Alyssa C Frazee, Ben Langmead, and Jeffrey T Leek. Recount: A multi-experiment resource of analysis-ready RNA-seq gene count datasets. *BMC Bioinformatics*, 12(1):449, 2011.

[38] Stacey B Gabriel, Stephen F Schaffner, Huy Nguyen, Jamie M Moore, Jessica Roy, Brendan Blumenstiel, John Higgins, Matthew DeFelice, Amy Lochner, Maura Faggart, et al. The structure of haplotype blocks in the human genome. *Science*, 296(5576):2225–2229, 2002.

[39] Bernat Gel, Anna Díez-Villanueva, Eduard Serra, Marcus Buschbeck, Miguel A Peinado, and Roberto Malinverni. regioner: an r/bioconductor package for the association analysis of genomic regions based on permutation tests. *Bioinformatics*, 32(2):289–291, 2015.

[40] Robert C Gentleman, Vincent J Carey, Douglas M Bates, Ben Bolstad, Marcel Dettling, Sandrine Dudoit, Byron Ellis, Laurent Gautier, Yongchao Ge, Jeff Gentry, et al. Bioconductor: Open software development for computational biology and bioinformatics. *Genome Biology*, 5(10):R80, 2004.

[41] J. R. González, I. Subirana, G. Escaramis, S. Peraza, A. Caceres, X. Estivill, and L. Armengol. Accounting for uncertainty when assessing association between copy number and disease: A latent class model. *BMC Bioinformatics*, 10:172, 2009.

[42] Juan R González, Lluís Armengol, Xavier Solé, Elisabet Guinó, Josep M Mercader, Xavier Estivill, and Víctor Moreno. SNPassoc: An R package to perform whole genome association studies. *Bioinformatics*, 23(5):654–655, 2007.

[43] Juan R González, Josep L Carrasco, Frank Dudbridge, Lluís Armengol, Xavier Estivill, and Victor Moreno. Maximizing association statistics over genetic models. *Genetic Epidemiology*, 32(3):246–254, 2008.

[44] Juan R González, Marta González-Carpio, Rosario Hernández-Sáez, Victoria Serrano Vargas, Guadalupe Torres Hidalgo, Marta Rubio-Rodrigo, Ana García-Nogales, Manuela Núñez Estévez, Luis M Luengo Pérez, and Raquel Rodríguez-López. FTO risk haplotype among early onset and severe obesity cases in a population of western spain. *Obesity*, 20(4):909–915, 2012.

[45] Juan R González, Benjamín Rodríguez-Santiago, Alejandro Cáceres, Roger Pique-Regi, Nathaniel Rothman, Stephen J Chanock, Lluís Armengol, and Luis A Pérez-Jurado. A fast and accurate method to detect allelic genomic imbalances underlying mosaic rearrangements using SNP array data. *BMC Bioinformatics*, 12(1):166, 2011.

[46] Jian Gu, Dong Liang, Yunfei Wang, Charles Lu, and Xifeng Wu. Effects of n-acetyl transferase 1 and 2 polymorphisms on bladder cancer risk in caucasians. *Mutation Research/Genetic Toxicology and Environmental Mutagenesis*, 581(1):97–104, 2005.

[47] Monica Guxens, Ferran Ballester, Mercedes Espada, Mariana F Fernández, Joan O Grimalt, Jesús Ibarluzea, Nicolás Olea, Marisa Rebagliato, Adonina Tardón, Maties Torrent, et al. Cohort profile: The INMAiNfancia y Medio Ambiente(environment and childhood) project. *International Journal of Epidemiology*, 41(4):930–940, 2011.

[48] Florian Hahne and Robert Ivanek. Visualizing genomic data using Gviz and Bioconductor. *Statistical Genomics: Methods and protocols*, pages 335–351, 2016.

[49] Saskia Haitjema, Daniel Kofink, Jessica Van Setten, Sander W Van Der Laan, Arjan H Schoneveld, James Eales, Maciej Tomaszewski, Saskia CA De Jager, Gerard Pasterkamp, Folkert W Asselbergs, et al. Loss of y chromosome in blood is associated with major cardiovascular events during follow-up in men after carotid endarterectomy. *Circulation: Genomic and Precision Medicine*, 10(4):e001544, 2017.

[50] Kasper D Hansen, Rafael A Irizarry, and Zhijin Wu. Removing technical variability in RNA-seq data using conditional quantile normalization. *Biostatistics*, 13(2):204–216, 2012.

[51] Kasper Daniel Hansen, Winston Timp, Héctor Corrada Bravo, Sarven Sabunciyan, Benjamin Langmead, Oliver G McDonald, Bo Wen, Hao Wu, Yun Liu, Dinh Diep, et al. Increased methylation variation in epigenetic domains across cancer types. *Nature Genetics*, 43(8):768, 2011.

[52] Carles Hernandez-Ferrer, Ines Quintela Garcia, Katharina Danielski, Ángel Carracedo, Luis A Pérez-Jurado, and Juan R González. affy2sv: An R package to pre-process Affymetrix Cytoscan HD and 750k arrays for SNP, CNV, inversion and mosaicism calling. *BMC Bioinformatics*, 16(1):167, 2015.

[53] Carles Hernandez-Ferrer and Juan R González. CTDquerier: A Bioconductor R package for comparative toxicogenomics databaseTM data extraction, visualization and enrichment of environmental and toxicological studies. *Bioinformatics*, 1:3, 2018.

[54] Carles Hernandez-Ferrer, Carlos Ruiz-Arenas, Alba Beltran-Gomila, and Juan R González. Multidataset: An R package for encapsulating multiple data sets with application to omic data integration. *BMC Bioinformatics*, 18(1):36, 2017.

[55] Lucia A Hindorff, Elizabeth M Gillanders, and Teri A Manolio. Genetic architecture of cancer and other complex diseases: Lessons learned and future directions. *Carcinogenesis*, 32(7):945–954, 2011.

[56] E Andres Houseman, Molly L Kile, David C Christiani, Tan A Ince, Karl T Kelsey, and Carmen J Marsit. Reference-free deconvolution of DNA methylation data and mediation by cell composition effects. *BMC Bioinformatics*, 17(1):259, 2016.

[57] Eugene Andres Houseman, William P Accomando, Devin C Koestler, Brock C Christensen, Carmen J Marsit, Heather H Nelson, John K Wiencke, and Karl T Kelsey. DNA methylation arrays as surrogate measures of cell mixture distribution. *BMC Bioinformatics*, 13(1):86, 2012.

[58] Eugene Andres Houseman, John Molitor, and Carmen J Marsit. Reference-free cell mixture adjustments in analysis of DNA methylation data. *Bioinformatics*, 30(10):1431–1439, 2014.

[59] Bryan Howie, Christian Fuchsberger, Matthew Stephens, Jonathan Marchini, and Gonçalo R Abecasis. Fast and accurate genotype imputation in genome-wide association studies through pre-phasing. *Nature Genetics*, 44(8):955–959, 2012.

[60] Da Wei Huang, Brad T Sherman, and Richard A Lempicki. Bioinformatics enrichment tools: paths toward the comprehensive functional analysis of large gene lists. *Nucleic acids research*, 37(1):1–13, 2008.

[61] Da Wei Huang, Brad T Sherman, and Richard A Lempicki. Systematic and integrative analysis of large gene lists using david bioinformatics resources. *Nature protocols*, 4(1):44, 2008.

[62] Yen-Tsung Huang, Tyler J VanderWeele, and Xihong Lin. Joint analysis of SNP and gene expression data in genetic association studies of complex diseases. *The Annals of Applied Statistics*, 8(1):352, 2014.

[63] Wolfgang Huber, Vincent J Carey, Robert Gentleman, Simon Anders, Marc Carlson, Benilton S Carvalho, Hector Corrada Bravo, Sean Davis, Laurent Gatto, Thomas Girke, et al. Orchestrating high-throughput genomic analysis with Bioconductor. *Nature Methods*, 12(2):115–121, 2015.

[64] Kevin B Jacobs, Meredith Yeager, Weiyin Zhou, Sholom Wacholder, Zhaoming Wang, Benjamin Rodriguez-Santiago, Amy Hutchinson, Xiang Deng, Chenwei Liu, Marie-Josephe Horner, et al. Detectable clonal mosaicism and its relationship to aging and cancer. *Nature Genetics*, 44(6):651, 2012.

[65] Rudolf Jaenisch and Adrian Bird. Epigenetic regulation of gene expression: How the genome integrates intrinsic and environmental signals. *Nature Genetics*, 33:245–254, 2003.

[66] W Evan Johnson, Cheng Li, and Ariel Rabinovic. Adjusting batch effects in microarray expression data using empirical Bayes methods. *Biostatistics*, 8(1):118–127, 2007.

[67] Jungnam Joo, Minjung Kwak, Zehua Chen, and Gang Zheng. Efficiency robust statistics for genetic linkage and association studies under genetic model uncertainty. *Statistics in Medicine*, 29(1):158–180, 2010.

[68] Minoru Kanehisa and Susumu Goto. Kegg: Kyoto encyclopedia of genes and genomes. *Nucleic Acids Research*, 28(1):27–30, 2000.

[69] Purvesh Khatri, Marina Sirota, and Atul J Butte. Ten years of pathway analysis: current approaches and outstanding challenges. *PLoS computational biology*, 8(2):e1002375, 2012.

[70] Daniel A King, Wendy D Jones, Yanick J Crow, Anna F Dominiczak, Nicola A Foster, Tom R Gaunt, Jade Harris, Stephen W Hellens, Tessa Homfray, Josie Innes, et al. Mosaic structural variation in children with developmental disorders. *Human Molecular Genetics*, 24(10):2733–2745, 2015.

[71] Mark Kirkpatrick and Nick Barton. Chromosome inversions, local adaptation and speciation. *Genetics*, 173(1):419–434, 2006.

[72] J. M. Korn, F. G. Kuruvilla, S. A. McCarroll, A. Wysoker, J. Nemesh, S. Cawley, E. Hubbell, J. Veitch, P. J. Collins, K. Darvishi, C. Lee, M. M. Nizzari, S. B. Gabriel, S. Purcell, M. J. Daly, and D. Altshuler. Integrated genotype calling and association analysis of SNPs, common copy number polymorphisms and rare CNVs. *Nature Genetics*, 40(10):1253–60, 2008.

[73] Jean-Charles Lambert, Carla A Ibrahim-Verbaas, Denise Harold, Adam C Naj, Rebecca Sims, Céline Bellenguez, Gyungah Jun, Anita L DeStefano, Joshua C Bis, Gary W Beecham, et al. Meta-analysis of 74,046 individuals identifies 11 new susceptibility loci for Alzheimer's disease. *Nature Genetics*, 45(12):1452–1458, 2013.

[74] Ilkka Lappalainen, Jeff Almeida-King, Vasudev Kumanduri, Alexander Senf, John Dylan Spalding, Gary Saunders, Jag Kandasamy, Mario Caccamo, Rasko Leinonen, Brendan Vaughan, et al. The european genome-phenome archive of human data consented for biomedical research. *Nature genetics*, 47(7):692, 2015.

[75] Charity W Law, Yunshun Chen, Wei Shi, and Gordon K Smyth. Voom: Precision weights unlock linear model analysis tools for RNA-seq read counts. *Genome Biology*, 15(2):R29, 2014.

[76] Michael Lawrence, Wolfgang Huber, Hervé Pages, Patrick Aboyoun, Marc Carlson, Robert Gentleman, Martin T Morgan, and Vincent J Carey. Software for computing and annotating genomic ranges. *PLoS Computational Biology*, 9(8):e1003118, 2013.

[77] Joseph H Lee, Rong Cheng, Neill Graff-Radford, Tatiana Foroud, and Richard Mayeux. Analyses of the national institute on aging late-onset Alzheimer's disease family study: Implication of additional loci. *Archives of Neurology*, 65(11):1518–1526, 2008.

[78] Jeffrey T Leek, W Evan Johnson, Hilary S Parker, Andrew E Jaffe, and John D Storey. The SVA package for removing batch effects and other unwanted variation in high-throughput experiments. *Bioinformatics*, 28(6):882–883, 2012.

[79] Jeffrey T Leek, Robert B Scharpf, Héctor Corrada Bravo, David Simcha, Benjamin Langmead, W Evan Johnson, Donald Geman, Keith Baggerly, and Rafael A Irizarry. Tackling the widespread and critical impact of batch effects in high-throughput data. *Nature Reviews Genetics*, 11(10):733, 2010.

[80] Guillaume Lettre, Christoph Lange, and Joel N Hirschhorn. Genetic model testing and statistical power in population-based association studies of quantitative traits. *Genetic Epidemiology*, 31(4):358–362, 2007.

[81] Jie-Qiong Li, Jin-Tai Yu, Teng Jiang, and Lan Tan. Endoplasmic reticulum dysfunction in Alzheimers disease. *Molecular Neurobiology*, 51(1):383–395, 2015.

[82] Miao-Xin Li, Juilian MY Yeung, Stacey S Cherny, and Pak C Sham. Evaluating the effective numbers of independent tests and significant p-value thresholds in commercial genotyping arrays and public imputation reference datasets. *Human Genetics*, 131(5):747–756, 2012.

[83] Xiaohong Li, Guy N Brock, Eric C Rouchka, Nigel GF Cooper, Dongfeng Wu, Timothy E OToole, Ryan S Gill, Abdallah M Eteleeb, Liz OBrien, and Shesh N Rai. A comparison of per sample global scaling and per gene normalization methods for differential expression analysis of RNA-seq data. *PloS One*, 12(5):e0176185, 2017.

[84] Jeffrey C Long, Robert C Williams, and Margrit Urbanek. An EM algorithm and testing strategy for multiple-locus haplotypes. *American Journal of Human Genetics*, 56(3):799, 1995.

[85] John Lonsdale, Jeffrey Thomas, Mike Salvatore, Rebecca Phillips, Edmund Lo, Saboor Shad, Richard Hasz, Gary Walters, Fernando Garcia, Nancy Young, et al. The genotype-tissue expression (GTEx) project. *Nature Genetics*, 45(6):580–585, 2013.

[86] Jianzhong Ma and Christopher I Amos. Investigation of inversion polymorphisms in the human genome using principal components analysis. *PloS One*, 7(7):e40224, 2012.

[87] Jacqueline MacArthur, Emily Bowler, Maria Cerezo, Laurent Gil, Peggy Hall, Emma Hastings, Heather Junkins, Aoife McMahon, Annalisa Milano, Joannella Morales, et al. The new NHGRI-EBI catalog of published genome-wide association studies (GWAS catalog). *Nucleic Acids Research*, 45(D1):D896–D901, 2016.

[88] Matthew D Mailman, Michael Feolo, Yumi Jin, Masato Kimura, Kimberly Tryka, Rinat Bagoutdinov, Luning Hao, Anne Kiang, Justin Paschall, Lon Phan, et al. The NCBI dbGaP database of genotypes and phenotypes. *Nature Genetics*, 39(10):1181–1186, 2007.

[89] Teri A Manolio. In retrospect: A decade of shared genomic associations. *Nature*, 546(7658):360–361, 2017.

[90] Teri A Manolio, Francis S Collins, Nancy J Cox, David B Goldstein, Lucia A Hindorff, David J Hunter, Mark I McCarthy, Erin M Ramos,

Lon R Cardon, Aravinda Chakravarti, et al. Finding the missing heritability of complex diseases. *Nature*, 461(7265):747–753, 2009.

[91] John C Marioni, Christopher E Mason, Shrikant M Mane, Matthew Stephens, and Yoav Gilad. Rna-seq: An assessment of technical reproducibility and comparison with gene expression arrays. *Genome Research*, 18(9):1509–1517, 2008.

[92] Chen Meng, Bernhard Kuster, Aedín C Culhane, and Amin Moghaddas Gholami. A multivariate approach to the integration of multi-omics datasets. *BMC Bioinformatics*, 15(1):162, 2014.

[93] Chen Meng, Oana A Zeleznik, Gerhard G Thallinger, Bernhard Kuster, Amin M Gholami, and Aedín C Culhane. Dimension reduction techniques for the integrative analysis of multi-omics data. *Briefings in Bioinformatics*, 17(4):628–641, 2016.

[94] J Mill, S Richards, Jo Knight, S Curran, Eleanor Taylor, and Philip Asherson. Haplotype analysis of SNAP-25 suggests a role in the aetiology of adhd. *Molecular Psychiatry*, 9(8):801, 2004.

[95] Joshua Millstein, Gary K Chen, and Carrie V Breton. Cit: Hypothesis testing software for mediation analysis in genomic applications. *Bioinformatics*, 32(15):2364–2365, 2016.

[96] Joshua Millstein, Bin Zhang, Jun Zhu, and Eric E Schadt. Disentangling molecular relationships with a causal inference test. *BMC Genetics*, 10(1):23, 2009.

[97] JL Min, G Hemani, G Davey Smith, C Relton, M Suderman, and John Hancock. Meffil: Efficient normalization and analysis of very large DNA methylation datasets. *Bioinformatics*, 2018.

[98] Ali Mortazavi, Brian A Williams, Kenneth McCue, Lorian Schaeffer, and Barbara Wold. Mapping and quantifying mammalian transcriptomes by RNA-seq. *Nature Methods*, 5(7):621, 2008.

[99] Myocardial Infarction Genetics Consortium. Genome-wide association of early-onset myocardial infarction with single nucleotide polymorphisms and copy number variants. *Nature Genetics*, 41(3):334–341, 2009.

[100] Rajan P Nair, Philip E Stuart, Ioana Nistor, Ravi Hiremagalore, Nicholas VC Chia, Stefan Jenisch, Michael Weichenthal, Gonçalo R Abecasis, Henry W Lim, Enno Christophers, et al. Sequence and haplotype analysis supports HLA-C as the psoriasis susceptibility 1 gene. *The American Journal of Human Genetics*, 78(5):827–851, 2006.

[101] Cancer Genome Atlas Network et al. Comprehensive molecular portraits of human breast tumours. *Nature*, 490(7418):61–70, 2012.

[102] R. M. Neve, K. Chin, J. Fridlyand, J. Yeh, F. L. Baehner, T. Fevr, L. Clark, N Bayani, J-P. Coppe, F. Tong, T. Speed, P. T. Spellman, S. DeVries, A. Lapuk, N. J. Wang, W-L. Kuo, J. L. Stilwell, D. Pinkel, D. G. Albertson, F. M. Waldman, F. McCormick, R. B. Dickson, M. D. Johnson, M. Lippman, S. Ethier, A. Gazdar, and J. W. Gray. A collection of breast cancer cell lines for the study of functionally distinct cancer subtypes. *Cancer Cell*, 10:515 – 527, 2006.

[103] Dan L Nicolae, Eric Gamazon, Wei Zhang, Shiwei Duan, M Eileen Dolan, and Nancy J Cox. Trait-associated SNPs are more likely to be eQTLs: Annotation to enhance discovery from GWAS. *PLoS genetics*, 6(4):e1000888, 2010.

[104] Gro Nilsen, Knut Liestøl, Peter Van Loo, Hans Kristian Moen Vollan, Marianne B Eide, Oscar M Rueda, Suet-Feung Chin, Roslin Russell, Lars O Baumbusch, Carlos Caldas, et al. Copynumber: Efficient algorithms for single-and multi-track copy number segmentation. *BMC Genomics*, 13(1):591, 2012.

[105] Adam B Olshen, ES Venkatraman, Robert Lucito, and Michael Wigler. Circular binary segmentation for the analysis of array-based DNA copy number data. *Biostatistics*, 5(4):557–572, 2004.

[106] Chirag J Patel, Jayanta Bhattacharya, and Atul J Butte. An environment-wide association study (EWAS) on type 2 diabetes mellitus. *PloS One*, 5(5):e10746, 2010.

[107] Chirag J Patel, Nam Pho, Michael McDuffie, Jeremy Easton-Marks, Cartik Kothari, Isaac S Kohane, and Paul Avillach. A database of human exposomes and phenomes from the US national health and nutrition examination survey. *Scientific Data*, 3:160096, 2016.

[108] Itsik Pe'er, Roman Yelensky, David Altshuler, and Mark J Daly. Estimation of the multiple testing burden for genomewide association studies of nearly all common variants. *Genetic epidemiology*, 32(4):381–385, 2008.

[109] F. Picard, S. Robin, E. Lebarbier, and J. J. Daudin. A segmentation/clustering model for the analysis of array CGH data. *Biometrics*, 63(3):758–766, 2007.

[110] Joseph K Pickrell, John C Marioni, Athma A Pai, Jacob F Degner, Barbara E Engelhardt, Everlyne Nkadori, Jean-Baptiste Veyrieras, Matthew Stephens, Yoav Gilad, and Jonathan K Pritchard. Understanding mechanisms underlying human gene expression variation with RNA sequencing. *Nature*, 464(7289):768, 2010.

[111] Roger Pique-Regi, Alejandro Cáceres, and Juan R González. R-gada: A fast and flexible pipeline for copy number analysis in association studies. *BMC Bioinformatics*, 11(1):380, 2010.

[112] Roger Pique-Regi, Jordi Monso-Varona, Antonio Ortega, Robert C Seeger, Timothy J Triche, and Shahab Asgharzadeh. Sparse representation and Bayesian detection of genome copy number alterations from microarray data. *Bioinformatics*, 24(3):309–318, 2008.

[113] Alkes L Price, Nick J Patterson, Robert M Plenge, Michael E Weinblatt, Nancy A Shadick, and David Reich. Principal components analysis corrects for stratification in genome-wide association studies. *Nature Genetics*, 38(8):904, 2006.

[114] Alkes L Price, Noah A Zaitlen, David Reich, and Nick Patterson. New approaches to population stratification in genome-wide association studies. *Nature Reviews Genetics*, 11(7):459, 2010.

[115] Shaun Purcell, Benjamin Neale, Kathe Todd-Brown, Lori Thomas, Manuel AR Ferreira, David Bender, Julian Maller, Pamela Sklar, Paul IW De Bakker, Mark J Daly, et al. Plink: A tool set for whole-genome association and population-based linkage analyses. *The American Journal of Human Genetics*, 81(3):559–575, 2007.

[116] Johannes Rainer. *EnsDb.Hsapiens.v86: Ensembl based annotation package*, 2017. R package version 2.99.0.

[117] Marcel Ramos, Lucas Schiffer, Angela Re, Rimsha Azhar, Azfar Basunia, Carmen Rodriguez, Tiffany Chan, Phil Chapman, Sean R Davis, David Gomez-Cabrero, et al. Software for the integration of multiomics experiments in Bioconductor. *Cancer Research*, 77(21):e39–e42, 2017.

[118] Stephen M Rappaport and Martyn T Smith. Environment and disease risks. *Science*, 330(6003):460–461, 2010.

[119] Rahul Reddy. A comparison of methods: Normalizing high-throughput RNA sequencing data. *bioRxiv*, 2015.

[120] Judith Reina-Castillón, Roser Pujol, Marcos López-Sánchez, Benjamín Rodríguez-Santiago, Miriam Aza-Carmona, Juan Ramón González, José Antonio Casado, Juan Antonio Bueren, Julián Sevilla, Isabel Badel, et al. Detectable clonal mosaicism in blood as a biomarker of cancer risk in Fanconi anemia. *Blood Advances*, 1(5):319–329, 2017.

[121] A. Reiner, D. Yekutieli, and Y Benjamini. Identifying differentially expressed genes using false discovery rate controlling procedures. *Bioinformatics*, 19(3):368–375, 2003.

[122] RC Richmond, G Hemani, K Tilling, G Davey Smith, and CL Relton. Challenges and novel approaches for investigating molecular mediation. *Human Molecular Genetics*, 25(R2):R149–R156, 2016.

[123] Marylyn D Ritchie, Emily R Holzinger, Ruowang Li, Sarah A Pendergrass, and Dokyoon Kim. Methods of integrating data to uncover genotype–phenotype interactions. *Nature Reviews Genetics*, 16(2):85, 2015.

[124] Mark D Robinson, Davis J McCarthy, and Gordon K Smyth. edger: A Bioconductor package for differential expression analysis of digital gene expression data. *Bioinformatics*, 26(1):139–140, 2010.

[125] Mark D Robinson and Alicia Oshlack. A scaling normalization method for differential expression analysis of RNA-seq data. *Genome Biology*, 11(3):R25, 2010.

[126] Benjamín Rodríguez-Santiago, Nuria Malats, Nathaniel Rothman, Lluís Armengol, Montse Garcia-Closas, Manolis Kogevinas, Olaya Villa, Amy Hutchinson, Julie Earl, Gaëlle Marenne, et al. Mosaic uniparental disomies and aneuploidies as large structural variants of the human genome. *The American Journal of Human Genetics*, 87(1):129–138, 2010.

[127] Nathaniel Rothman, Montserrat Garcia-Closas, Nilanjan Chatterjee, Nuria Malats, Xifeng Wu, Jonine D Figueroa, Francisco X Real, David Van Den Berg, Giuseppe Matullo, Dalsu Baris, et al. A multi-stage genome-wide association study of bladder cancer identifies multiple susceptibility loci. *Nature Genetics*, 42(11):978, 2010.

[128] Carlos Ruiz and Juan R. Gonzalez. *scoreInvHap: Get inversion status in predefined regions*. R package version 1.5.2.

[129] Carlos Ruiz-Arenas and Juan R González. Redundancy analysis allows improved detection of methylation changes in large genomic regions. *BMC Bioinformatics*, 18(1):553, 2017.

[130] Carlos Ruiz-Arenas, Carles Hernandez-Ferrer, and Juan R. González. *MEAL: Perform methylation analysis*, 2017. R package version 1.6.0.

[131] Peter Rzehak, Marcela Covic, Richard Saffery, Eva Reischl, Simone Wahl, Veit Grote, Martina Weber, Annick Xhonneux, Jean-Paul Langhendries, Natalia Ferre, et al. DNA-methylation and body composition in preschool children: Epigenome-wide-analysis in the European Childhood Obesity Project (CHOP)-study. *Scientific Reports*, 7(1):14349, 2017.

[132] Nidhee M Sachdev, Susan M Maxwell, Andria G Besser, and James A Grifo. Diagnosis and clinical management of embryonic mosaicism. *Fertility and Sterility*, 107(1):6–11, 2017.

[133] Maximilian PA Salm, Stuart D Horswell, Claire E Hutchison, Helen E Speedy, Xia Yang, Liming Liang, Eric E Schadt, William O Cookson,

Anthony S Wierzbicki, Rossi P Naoumova, et al. The origin, global distribution, and functional impact of the human 8p23 inversion polymorphism. *Genome Research*, 22(6):1144–1153, 2012.

[134] Robert B Scharpf, Ingo Ruczinski, Benilton Carvalho, Betty Doan, Aravinda Chakravarti, and Rafael A Irizarry. A multilevel model to address batch effects in copy number estimation using SNP arrays. *Biostatistics*, 12(1):33–50, 2010.

[135] André Scherag, Christian Dina, Anke Hinney, Vincent Vatin, Susann Scherag, Carla IG Vogel, Timo D Müller, Harald Grallert, H-Erich Wichmann, Beverley Balkau, et al. Two new loci for body-weight regulation identified in a joint analysis of genome-wide association studies for early-onset extreme obesity in French and German study groups. *PLoS Genetics*, 6(4):e1000916, 2010.

[136] Stuart A Scott, Ninette Cohen, Tracy Brandt, Gokce Toruner, Robert J Desnick, and Lisa Edelmann. Detection of low-level mosaicism and placental mosaicism by oligonucleotide array comparative genomic hybridization. *Genetics in Medicine*, 12(2):85, 2010.

[137] J.-H. Shin, S. Blay, B. McNeney, and J. Graham. LDheatmap: An R function for graphical display of pairwise linkage disequilibria between single nucleotide polymorphisms. *Journal of Statistical Software*, 16:Code Snippet 3, 2006.

[138] Patrick E Shrout and Niall Bolger. Mediation in experimental and non-experimental studies: New procedures and recommendations. *Psychological Methods*, 7(4):422, 2002.

[139] Mark S Silverberg, Judy H Cho, John D Rioux, Dermot PB McGovern, Jing Wu, Vito Annese, Jean-Paul Achkar, Philippe Goyette, Regan Scott, Wei Xu, et al. Ulcerative colitis–risk loci on chromosomes 1p36 and 12q15 found by genome-wide association study. *Nature Genetics*, 41(2):216, 2009.

[140] Michael E Sobel. Asymptotic confidence intervals for indirect effects in structural equation models. *Sociological Methodology*, 13:290–312, 1982.

[141] Charlotte Soneson and Mauro Delorenzi. A comparison of methods for differential expression analysis of RNA-seq data. *BMC Bioinformatics*, 14(1):91, 2013.

[142] Sanjana Sood, Iain J Gallagher, Katie Lunnon, Eric Rullman, Aoife Keohane, Hannah Crossland, Bethan E Phillips, Tommy Cederholm, Thomas Jensen, Luc JC van Loon, et al. A novel multi-tissue RNA diagnostic of healthy ageing relates to cognitive health status. *Genome Biology*, 16(1):185, 2015.

[143] Rebecca V Steenaard, Symen Ligthart, Lisette Stolk, Marjolein J Peters, Joyce B van Meurs, Andre G Uitterlinden, Albert Hofman, Oscar H Franco, and Abbas Dehghan. Tobacco smoking is associated with methylation of genes related to coronary artery disease. *Clinical Epigenetics*, 7(1):54, 2015.

[144] Hreinn Stefansson, Agnar Helgason, Gudmar Thorleifsson, Valgerdur Steinthorsdottir, Gisli Masson, John Barnard, Adam Baker, Aslaug Jonasdottir, Andres Ingason, Vala G Gudnadottir, et al. A common inversion under selection in Europeans. *Nature Genetics*, 37(2):129–137, 2005.

[145] Matthew Stephens and Peter Donnelly. A comparison of Bayesian methods for haplotype reconstruction from population genotype data. *The American Journal of Human Genetics*, 73(5):1162–1169, 2003.

[146] Matthew Stephens, Nicholas J Smith, and Peter Donnelly. A new statistical method for haplotype reconstruction from population data. *The American Journal of Human Genetics*, 68(4):978–989, 2001.

[147] Victoria Stodden, Friedrich Leisch, and Roger D Peng. *Implementing Reproducible Research*. CRC Press, 2014.

[148] AH Sturtevant. A case of rearrangement of genes in drosophila. *Proceedings of the National Academy of Sciences*, 7(8):235–237, 1921.

[149] Aravind Subramanian, Pablo Tamayo, Vamsi K Mootha, Sayan Mukherjee, Benjamin L Ebert, Michael A Gillette, Amanda Paulovich, Scott L Pomeroy, Todd R Golub, Eric S Lander, et al. Gene set enrichment analysis: A knowledge-based approach for interpreting genome-wide expression profiles. *Proceedings of the National Academy of Sciences*, 102(43):15545–15550, 2005.

[150] Terje Svingen and Anne Marie Vinggaard. The risk of chemical cocktail effects and how to deal with the issue, 2015.

[151] Moshe Szyf and Johanna Bick. DNA methylation: A mechanism for embedding early life experiences in the genome. *Child Development*, 84(1):49–57, 2013.

[152] Arthur Tenenhaus, Cathy Philippe, Vincent Guillemot, Kim-Anh Le Cao, Jacques Grill, and Vincent Frouin. Variable selection for generalized canonical correlation analysis. *Biostatistics*, 15(3):569–583, 2014.

[153] Arthur Tenenhaus and Michel Tenenhaus. Regularized generalized canonical correlation analysis. *Psychometrika*, 76(2):257, 2011.

[154] Michel Tenenhaus, Arthur Tenenhaus, and Patrick JF Groenen. Regularized generalized canonical correlation analysis: A framework for sequential multiblock component methods. *Psychometrika*, 82(3):737–777, 2017.

[155] Dustin Tingley, Teppei Yamamoto, Kentaro Hirose, Luke Keele, and Kosuke Imai. Mediation: R package for causal mediation analysis. *Journal of Statistical Software*, 59:1, 2014.

[156] Katarzyna Tomczak, Patrycja Czerwińska, and Maciej Wiznerowicz. The cancer genome atlas (TCGA): An immeasurable source of knowledge. *Contemporary Oncology*, 19(1A):A68, 2015.

[157] Akira Toriba, Thaneeya Chetiyanukornkul, Ryoichi Kizu, and Kazuichi Hayakawa. Quantification of 2-hydroxyfluorene in human urine by column-switching high performance liquid chromatography with fluorescence detection. *Analyst*, 128(6):605–610, 2003.

[158] Cole Trapnell, Brian A Williams, Geo Pertea, Ali Mortazavi, Gordon Kwan, Marijke J Van Baren, Steven L Salzberg, Barbara J Wold, and Lior Pachter. Transcript assembly and quantification by RNA-seq reveals unannotated transcripts and isoform switching during cell differentiation. *Nature Biotechnology*, 28(5):511, 2010.

[159] Sebastian Treusch, Shusei Hamamichi, Jessica L Goodman, Kent ES Matlack, Chee Yeun Chung, Valeriya Baru, Joshua M Shulman, Antonio Parrado, Brooke J Bevis, Julie S Valastyan, et al. Functional links between aβ toxicity, endocytic trafficking, and Alzheimers disease risk factors in yeast. *Science*, 334(6060):1241–1245, 2011.

[160] Monica Uddin, Allison E Aiello, Derek E Wildman, Karestan C Koenen, Graham Pawelec, Regina de Los Santos, Emily Goldmann, and Sandro Galea. Epigenetic and immune function profiles associated with posttraumatic stress disorder. *Proceedings of the National Academy of Sciences*, 107(20):9470–9475, 2010.

[161] Jodie B Ullman and Peter M Bentler. *Structural equation modeling*, volume 2. Wiley Online Library, 2012.

[162] M. A. van de Wiel, K. I. Kim, S. J. Vosse, W. N. van Wieringen, S. M. Wilting, and B. Ylstra. CGHcall: Calling aberrations for array CGH tumor profiles. *Bioinformatics*, 23(7):892–894, 2007.

[163] M. A. van de Wiel and W. N. van Wieringen. CGHregions: Dimension reduction for array CGH data with minimal information loss. *Cancer Informatics*, 2:55–63, 2007.

[164] Maarten van Iterson, Erik W van Zwet, and Bastiaan T Heijmans. Controlling bias and inflation in epigenome-and transcriptome-wide association studies using the empirical null distribution. *Genome Biology*, 18(1):19, 2017.

[165] Tyler J VanderWeele. Mediation analysis: A practitioner's guide. *Annual Review of Public Health*, 37:17–32, 2016.

[166] Martine Vrijheid, Rémy Slama, Oliver Robinson, Leda Chatzi, Muireann Coen, Peter Van den Hazel, Cathrine Thomsen, John Wright, Toby J Athersuch, Narcis Avellana, et al. The human early-life exposome (helix): Project rationale and design. *Environmental Health Perspectives*, 122(6):535, 2014.

[167] Kai Wang, Mingyao Li, Dexter Hadley, Rui Liu, Joseph Glessner, Struan FA Grant, Hakon Hakonarson, and Maja Bucan. PennCNV: An integrated hidden Markov model designed for high-resolution copy number variation detection in whole-genome SNP genotyping data. *Genome Research*, 17(11):1665–1674, 2007.

[168] Danielle Welter, Jacqueline MacArthur, Joannella Morales, Tony Burdett, Peggy Hall, Heather Junkins, Alan Klemm, Paul Flicek, Teri Manolio, Lucia Hindorff, et al. The nhgri GWAS catalog, a curated resource of SNP-trait associations. *Nucleic Acids Research*, 42(D1):D1001–D1006, 2013.

[169] Janis E Wigginton, David J Cutler, and Gonçalo R Abecasis. A note on exact tests of Hardy-Weinberg equilibrium. *The American Journal of Human Genetics*, 76(5):887–893, 2005.

[170] Christopher Paul Wild. Complementing the genome with an exposome: The outstanding challenge of environmental exposure measurement in molecular epidemiology, 2005.

[171] Christopher Paul Wild. The exposome: From concept to utility. *International Journal of Epidemiology*, 41(1):24–32, 2012.

[172] Daniela M Witten, Robert Tibshirani, and Trevor Hastie. A penalized matrix decomposition, with applications to sparse principal components and canonical correlation analysis. *Biostatistics*, 10(3):515–534, 2009.

[173] Daniela M Witten and Robert J Tibshirani. Extensions of sparse canonical correlation analysis with applications to genomic data. *Statistical Applications in Genetics and Molecular Biology*, 8(1):1–27, 2009.

[174] LF Wockner, EP Noble, BR Lawford, R McD Young, CP Morris, VLJ Whitehall, and J Voisey. Genome-wide DNA methylation analysis of human brain tissue from schizophrenia patients. *Translational Psychiatry*, 4(1):e339, 2014.

[175] Dongyan Wu, Haitao Yang, Stacey J Winham, Yanina Natanzon, Devin C Koestler, Tiane Luo, Brooke L Fridley, Ellen L Goode, Yanbo Zhang, and Yuehua Cui. Mediation analysis of alcohol consumption, DNA methylation, and epithelial ovarian cancer. *Journal of Human Genetics*, page 1, 2018.

[176] Jian Yang, S Hong Lee, Michael E Goddard, and Peter M Visscher. Gcta: A tool for genome-wide complex trait analysis. *The American Journal of Human Genetics*, 88(1):76–82, 2011.

[177] Guangchuang Yu, Li-Gen Wang, Yanyan Han, and Qing-Yu He. clusterProfiler: An R package for comparing biological themes among gene clusters. *Omics: a Journal of Integrative Biology*, 16(5):284–287, 2012.

[178] Kui Zhang, Minghua Deng, Ting Chen, Michael S Waterman, and Fengzhu Sun. A dynamic programming algorithm for haplotype block partitioning. *Proceedings of the National Academy of Sciences*, 99(11):7335–7339, 2002.

[179] Hongyu Zhao, Ruth Pfeiffer, and Mitchell H Gail. Haplotype analysis in population genetics and association studies. *Pharmacogenomics*, 4(2):171–178, 2003.

[180] Peter Zill, Andreas Büttner, Wolfgang Eisenmenger, Hans-Jürgen Möller, Brigitta Bondy, and Manfred Ackenheil. Single nucleotide polymorphism and haplotype analysis of a novel tryptophan hydroxylase isoform (tph2) gene in suicide victims. *Biological Psychiatry*, 56(8):581–586, 2004.

[181] James Zou, Christoph Lippert, David Heckerman, Martin Aryee, and Jennifer Listgarten. Epigenome-wide association studies without the need for cell-type composition. *Nature Methods*, 11(3):309, 2014.

Index